Cambridge

Elements in Politics and Society in Latin America
edited by
Maria Victoria Murillo
Columbia University
Tulia G. Falleti
University of Pennsylvania
Juan Pablo Luna
The Pontifical Catholic University of Chile
Andrew Schrank
Brown University

THE DISTRIBUTIVE POLITICS OF ENVIRONMENTAL PROTECTION IN LATIN AMERICA AND THE CARIBBEAN

Isabella Alcañiz
University of Maryland
Ricardo A. Gutiérrez
National University of San Martin and CONICET

CAMBRIDGE
UNIVERSITY PRESS

CAMBRIDGE
UNIVERSITY PRESS

University Printing House, Cambridge CB2 8BS, United Kingdom

One Liberty Plaza, 20th Floor, New York, NY 10006, USA

477 Williamstown Road, Port Melbourne, VIC 3207, Australia

314–321, 3rd Floor, Plot 3, Splendor Forum, Jasola District Centre,
New Delhi – 110025, India

103 Penang Road, #05–06/07, Visioncrest Commercial, Singapore 238467

Cambridge University Press is part of the University of Cambridge.

It furthers the University's mission by disseminating knowledge in the pursuit of
education, learning, and research at the highest international levels of excellence.

www.cambridge.org
Information on this title: www.cambridge.org/9781009263436
DOI: 10.1017/9781009263429

First published 2022

A catalogue record for this publication is available from the British Library.

ISBN 978-1-009-26343-6 Paperback
ISSN 2515-5253 (online)
ISSN 2515-5245 (print)

The Distributive Politics of Environmental Protection in Latin America and the Caribbean

Elements in Politics and Society in Latin America

DOI: 10.1017/9781009263429
First published online: August 2022

Isabella Alcañiz
University of Maryland

Ricardo A. Gutiérrez
National University of San Martin and CONICET

Author for correspondence: Isabella Alcañiz, ialcaniz@umd.edu

Abstract: The study of environmental politics in Latin America and the Caribbean expands as conflicts stemming from the deterioration of the natural world increase. Yet, this scholarship has not generated a broad research agenda similar to the ones that emerged around other key political phenomena. This Element seeks to address the lack of a comprehensive research agenda in Latin American and Caribbean environmental politics and helps integrate the existing, disparate literatures. Drawing from distributive politics, this Element asks who benefits from the appropriation and pollution of the environment, who pays the costs of climate change and environmental degradation, and who gains from the allocation of state protections.

Keywords: environmental politics, environmental protection, climate inequality, environmental justice, environmental conflict

ISBNs: 9781009263436 (PB), 9781009263429 (OC)
ISSNs: 2515-5253 (online), 2515-5245 (print)

Contents

1 Introduction

In Latin America and the Caribbean, environmental and climate politics are central to political life. The greatest opportunities and challenges faced by the region in recent years have been shaped by the regulation of natural resources and environmental conflict. Drivers of deep social change across Latin America and the Caribbean – like the commodities boom, the transformation of the energy matrix, indigenous contestation, the expansion of the agricultural frontier, and shifting population dynamics – have either stemmed from or resulted in increased costs for the natural world.

Research on environmental politics in the region has grown exponentially in recent times and has attempted to address a variety of problems stemming from the deteriorating environment. However, this scholarship has not generated a broad research agenda similar to the ones that emerged around other key political phenomena, such as regime transitions, the nature of political institutions, or economic voting.

This Element seeks to address the lack of a comprehensive research agenda in Latin American and Caribbean environmental politics (LACEP) and helps integrate the existing, disparate literatures. The lack of a comprehensive research agenda is partly due to the nature of environmental challenges, which vary greatly and generate different, seemingly unrelated problems to be studied – such as deforestation, water scarcity, and climate change. But perhaps more importantly, scholars' interpretations of these problems diverge widely. Empirical studies of environmental politics in LACEP tend to revolve around two separate paradigms where the locus of change is found in either mobilized society or the state. These two major approaches have dominated LACEP. However, as we show in the following sections, a number of studies and scholars link social actors to the state and bridge the two literatures. Even so, in this Element, we leverage this natural division of labor in order to facilitate the organization of our review. Thus, we caution the reader that some of the scholarship overlaps these two approaches and is discussed in both society- and state-focused sections. Conversely, given our choice of state–society divide, we omit some discussion of the governance and common pool resources-centered literature.

The seeming disconnection between scholars of social mobilization and the state in LACEP is the result of different questions being asked. On the one hand, the literature centered on the mobilization of social groups vying for environmental protection mostly inquires about the conditions under which actors forge collective action (Baver and Lynch, 2006a; Carruthers, 2008a; Christen et al., 1998; Foyer and Dumoulin Kevran, 2017; Lewis, 2016). It offers less information on the conditions under which protests may lead to state intervention.

While the state is necessarily involved as the sanctioning institution of environmental protection, this literature often overlooks the many links between state and society in the environmental regulatory process. On the other hand, the literature focused on state actions frequently examines policymaking as a segmented process, sometimes paying little attention to the demands that trigger it (Aguilar-Støen, 2018; Alcañiz and Gutiérrez, 2020a; Sears and Pinedo-Vasquez, 2011; Steinberg, 2001; Tecklin et al., 2011). While state-centered studies may consider some of the direct societal ties that shape state behavior, they sometimes neglect the political mobilization behind these links (Lemos, 1998). Research questions focused on the state tend to ask instead about the institutional mechanisms and state capabilities that produce environmental protection and induce changes in environmental governance.

This Element helps unify the study of environmental politics in Latin America and the Caribbean around one broad research paradigm. We begin with the understanding that environmental politics revolves around state decisions. Political actors look for environmental problems to be addressed by the state. Building upon the literature on Environmental Justice (EJ), we argue that environmental politics is as much about the protection of the environment as it is about the distribution of environmental benefits and ills. The process of social grievance and state address is fundamentally distributive, whereby some reap the benefits from the appropriation and pollution of natural resources while others pay the costs of environmental deterioration (Martínez-Alier et al., 2010). That is why we ask LACEP central research questions: Who *benefits* from the appropriation and pollution of the environment, who *pays* the costs of climate change and environmental degradation (i.e., the depletion of resources plus pollution), who *benefits* from the allocation of state protections, and who *pays* the cost of it.

Inequality significantly shapes this distribution, as the rich can use private means to safeguard themselves from climate change and the deteriorating environment, while the poor must rely solely on state protection (Acselrad, 2008). Giving existing asymmetries of power, some studies on the environment and rights (Acselrad, 2010, 2008, 2006; Alimonda, 2008, 2006; Carruthers, 2008a; Falleti and Riofrancos, 2018; Martínez-Alier, 2004) often find evidence that large companies and the state are the winners of environmental politics, while the people directly affected by private or state projects and activities (either the poor or local populations) are always the losers as they cannot change the course of business and state decisions. Some scholars find evidence of historical participatory patterns that benefit powerful economic actors and the state over Indigenous and other affected communities (Falleti and Riofrancos, 2018).

However, we understand that the distribution of environmental costs and benefits, similar to the distribution of the costs and benefits of industrialization, may be complicated by different political conditions, institutions, and agencies. We believe that the winners and losers of climate change and environmental degradation cannot be known a priori. Drawing from the literature on distributive politics and some of the studies reviewed in this Element, we argue that the distributive outcomes of environmental politics depend on the interaction among an array of social, economic, and state actors. More specifically, we find that often studies show that given certain conditions, affected populations can change the course of state decisions. At times, communities become winners of environmental politics when broad protectionist coalitions are built, bringing together diverse social, state, and even economic actors. A distributive politics approach facilitates less predetermined expectations on positions and outcomes, identifies crosscutting coalitions, and calls for a more open and dynamic analysis of the interaction of all actors involved in environmental politics.

Defining Environmental Politics

Environmental politics refers to the policy preferences, interests, and values as well as the system incentives that shape the actions of social, economic, and state actors defining and contesting environmental policy. Centered on the politics of environmental policies (or the lack thereof), this definition allows us to pay attention to the evaluative and strategic drivers as well as the contextual conditions of political action. While values, interests, and policy preferences describe actors' attributes, system incentives include both the governance arrangements and the economic and political contexts that shape those attributes and the resulting political actions.

The politics of environmental policies revolve around how and why decisions about the environment and the use of natural resources are made and implemented (Bull and Aguilar-Støen, 2015). The core of environmental policies lies in the state regulation of environmental protection, which becomes the aim of environmental politics. Environmental protection involves the preservation of the environment and the prevention of likely hazards as well as the remediation of harms already done. As we will discuss further, our approach to environmental politics focuses on the distributive effects of both hazards and harms, and preservation and remediation. The presence and absence of environmental policies affect the distributive costs and benefits of different actors.

Distributive effects ultimately refer to the actors' values, interests, and preferences: which values, preferences, and interests are better served by environmental policies. Interests constitute the benefits that are obtained from a certain object or a certain action, while preferences involve a choice between several objects or actions. If our interest is to maximize profits or to contemplate the landscape, we can do both in different ways. For example, we may prefer to produce soybeans rather than raise pigs, or we may prefer to look at a lake rather than a desert.

Value, in turn, is what makes a thing worthy of being appreciated, desired, and sought after. It is the quality that we appreciate in a certain object – for example, nature or the environment – and that guides our actions and our relationship with that object. In his analysis of the early ecologists of the eighteenth and nineteenth centuries, for example, Worster (1977) distinguishes two competing perspectives that still resonate in contemporary environmental conflicts. "Arcadian" ecology gives nature an intrinsic value that must be respected and which leads us to live in harmony with it, recognizing ourselves as just one part of that organic whole that we call nature. The "Imperial" tradition, on the other hand, conceives of nature as a resource for the progress of humanity, which must be dominated and transformed by human beings for their own glory and benefit. Throughout the Element, we will alternately talk about values, interests, and preferences. But it is important to note that we understand these three terms to be strongly connected. As Habermas states, interests, the preferences derived from them, and values cannot be thought of separately as "interests are connected with ideas [values] that justify normative validity claims" (Habermas, 1996: 69).

Environmental and distributive effects, thus, are intimately connected. Environmental effects refer to the state of the environment: how environmental policies or the absence of them impact the current and future conditions of the environment. Yet environmental effects are measured and evaluated by the involved actors on the basis of their distinct values, preferences, and interests (Martínez-Alier, 2004). For example, from an "eco-efficiency" perspective, provincial authorities and business actors saw gold mining in Esquel, Argentina, as a beneficial activity that could be compatible with environmental protection if adequate technologies and controls were set up. In contrast, an array of local actors opposed open-pit mining under the view that any development project for the city had to be grounded on local, small-scale activities such as agriculture, forestry, and tourism, which in their perspective were imbued with a higher appreciation of the environment (Walter and Martínez-Alier, 2010). Similarly, the environmental effects of forest certification in the Amazon were seen in different ways by different actors (Fearnside, 2003). Those who believed in "sustainable management" and were interested in securing cattle pasture defended certification as the best way to protect biodiversity, while those committed to "untouched forests" saw certification and management as disastrous for biodiversity.

Environmental Actors

From a distributive perspective, there is no way to distinguish independently the effects of a given human activity or environmental policy from the interests and preferences of the involved actors. Different actors have distinct preferences regarding environmental protection, which in turn shape how they interpret policies and their effects (Fearnside, 2003; Göbel and Ulloa, 2014; Lewis, 2016; Martínez-Alier, 2007; Martínez-Alier et al., 2010). Three types of actors are dominant in LACEP: state (e.g., policymakers, environmental officials, and government scientists), social (e.g., nongovernmental organizations or NGOs, local activists, Indigenous and peasant groups), and economic actors (e.g., private firms, agricultural producers, and state-owned companies). While the distinction between state and nonstate actors is usually clear-cut, we differentiate social and economic actors based on their profit orientation. Economic actors are profit-oriented, whereas social actors are nonprofit-oriented and may include elite and voluntary organizations and grassroots groups.

Social and economic actors do not have monolithic interests and preferences. Nevertheless, general trends can be identified when both groups are compared across cases of environmental conflict. Social actors involved in environmental politics typically prefer new or higher regulation and protection standards. Economic actors typically prefer to avoid or reduce regulations and protection standards. Yet, some economic actors occasionally may favor higher environmental protection and regulation in order to secure their economic interests – such as small-scale farming or tourism – against competing activities – such as mining or industrial agriculture (Alvarado Merino, 2008; Christel, 2019; McCaffrey and Baver, 2006; Murillo and Mangonnet, 2013; Urkidi and Walter, 2011; Walter and Martínez-Alier, 2010). They may also prefer environmental regulation as a means to mitigate skyrocketing social unrest and judicial pressures (Amengual, 2016, 2013; McAllister, 2008). Similarly, even if they agree on the need for more regulation, different social groups may hold different values and preferences as regards the same issue, as shown by the so-called people in parks debate – that is, the debate over humans living in protected areas. While some organizations favor the full protection of areas without people, others see the creation of large protected areas for natural conservation as causing harm to native peoples living on or near those lands and as ultimately offering less protection for nature because they lack the support of local residents (Fearnside, 2003).

The policy preferences of state actors are even harder to determine than those of social and economic actors. While some state actors charged with the implementation of environmental policies are alleged to prefer more regulation and higher protection standards, others prefer to lower protection standards or to avoid regulation altogether in support of economic production. These opposing preferences reflect the paradoxical nature of the modern state. The state is a bureaucratic organization that depends on capital accumulation as the source wherefrom to extract taxes at the same time that it operates within a society imbued with social, political, and ideological conflicts and from which it must extract its legitimacy (O'Connor, 1998; Offe, 1992; Polanyi, 2001). The modern state is thus subject to an inherent contradiction between the logic of capital accumulation and the logic of democratic legitimacy, and it must simultaneously respond to the requirements of both. This contradiction permeates the advance of environmental regulation in Latin America and the Caribbean in a paradoxical way. In order to protect the environment, the state needs additional funding. In many of the Latin American and Caribbean countries, budgetary surpluses often come from the extraction of natural resources (Lewis, 2016).

Never were the internal contradictions of the state sharper than over the last commodity boom, especially between 2004 and 2014. During this time, most Latin American and Caribbean governments (regardless of their progressive or conservative ideology) passed new environmental regulations and invested in conservation policies while simultaneously benefiting from windfall profits and revenues from extractive activities and the export of commodities (Bull and Aguilar-Støen, 2015; Castro et al., 2016; Svampa, 2019; Zimmerer, 2011).

On the one hand, the contradictory nature of the modern state generates multiple orientations and conflicts among state agencies with disparate goals (Göbel and Ulloa, 2014; Harrison, 1996; Hochstetler and Keck, 2007; McAllister, 2008; Merlinsky, 2013a; Scheberle, 2005), which in turn are exacerbated by the struggle between the different levels of government (Fearnside, 2003; Gutiérrez, 2018; Urkidi and Walter, 2011; Walter and Urkidi, 2016). On the other hand, it creates opportunities for the mobilization and influence of social organizations concerned with environmental protection (Gutiérrez, 2018; Hochstetler and Keck, 2007). As we show throughout the Element, the variation in the policy preferences of state actors is crucial to understand the politics of environmental policy and its effects. To a large extent, the success and failure of environmental protection initiatives rest on the collaboration and forged coalitions between state and social actors (Amengual, 2013; Castro et al., 2016; Christen et al., 1998; Gutiérrez, 2020, 2018; Hochstetler and Keck, 2007; Kauffman and Terry, 2016; Lemos, 1998;

Lemos and Looye, 2003; McCaffrey and Baver, 2006; Steinberg, 2001; Walter and Urkidi, 2016).

The Distribution of Environmental Costs and Benefits

Environmental politics is not only about the protection of the environment (i.e., what is to be protected) but also about the distribution of environmental costs and benefits (i.e., from whom and for whom it is to be protected). All environmental struggles are grounded in distributive conflict and all environmental policies entail distributive effects, even if the LACEP literature does not always analyze them explicitly. Examining the United States and Great Britain, Dobson (1998) contends that environmental sustainability and social justice are not necessarily compatible objectives – if they are not framed as opposite goals. The distinction between conservationism and the environmental justice movement illustrates well the tensions between sustainability and social justice. Conservation policies may very well help preserve certain ecosystems without solving social inequality problems such as access to natural resources and means of livelihood, exposure to pollution, and protection from environmental risks and disasters (Sachs, 1999).

Nevertheless, abundant studies in LACEP show that the kind of environmentalism that has developed in the region combines environmental and social concerns (Acselrad, 2010; Baver and Lynch, 2006a; Carey, 2009; Carruthers, 2008a; Cartagena Cruz, 2017; Hochstetler and Keck, 2007; Martínez-Alier et al., 2016; Sedrez, 2009; Silva, 2016; Svampa, 2012; Viola, 1992). This literature emphasizes the intertwining of environmental degradation and social justice, such as the destruction of traditional ways of life, access to land, and local livelihoods (Carruthers, 2008b; da Rocha et al., 2018; Leguizamón, 2020; Martínez-Alier, 2008; Perez Guartambel, 2006; Rodríguez et al., 2015; Sundberg, 2008; Ulloa, 2017; Urkidi and Walter, 2011). This scholarship finds that the communities most affected by climate change and the destruction of the natural environment are not just impoverished ones, but critically ones defined by ethnic, racial, and gender lines (Ahmed et al., 2021; Alcañiz and Sanchez-Rivera, forthcoming; Bledsoe, 2019; Futemma et al., 2015; Mollett, 2014; Perry, 2009; Rodriguez-Díaz and Lewellen-Williams, 2020; Tormos-Aponte, 2018; Vélez et al., 2020; Vida, 2020).

Indeed, over the past decades and across much of the region, we see Indigenous actors directly engaging in environmental issues. The rise of different ethnic coalitions is especially salient in Mexico, Central America, and the Andean countries. The ethnic divide is central to explaining outcomes in environmental politics (Armesto et al., 2001; Azócar et al., 2005; Carruthers and Rodriguez, 2009; Gonzalez, 2021; Inoue and Moreira, 2017; Orta-Martínez et al., 2018; Paredes, 2018; Paredes and Kaulard, 2020; Pragier, 2019; Valladares and Boelens, 2017; Wong, 2018; Zaremberg and Wong, 2018). We believe the LACEP literature will continue to catch up with this reality. We anticipate and hope it will help increase the visibility of Afro-Latin Americans' environmental grievances, especially those of Afro-Caribbean actors, who are even more marginalized from the LACEP literature.

By making explicit the distributive nature of environmental politics, we reveal the interaction of key actors in conservation struggles – and their interests and preferences – as well as its influence on green policies and the protection of the environment. The quest for environmental protection, we maintain, entails important questions about the costs and benefits that actors face, which in turn are shaped by the preferences and interests of these actors. On the one hand, environmental politics involves the winners and losers of the appropriation and pollution of the environment: who *benefits* from and who *pays* the costs of harms and hazards. On the other hand, it refers to the winners and losers of environmental protection policies: who *benefits* from the preservation and remediation of the environment and who *pays* the cost of the policies designed and implemented to achieve them. Even if we understand that in the long run, everyone will either pay the cost of environmental degradation or benefit from greater protection, what is at stake in real-world environmental politics are the lifetime calculations and aspirations of the actors involved (Hornborg, 2009).

The distributive nature of environmental politics is dynamic. If effective, environmental policies should change the costs and benefits that actors face. For example, one would expect the losers of environmental degradation to become the winners of environmental protection. Furthermore, we would also expect the winners of environmental degradation to bear at least some of the burden of remediation as well as the opportunity costs of restricting or banning a given polluting or degrading activity. This is why we pay special attention to the changing distributive effects of environmental degradation and policies in our survey of the LACEP literature.

This Element accomplishes two goals. First, Sections 2 and 3 offer compre-hensive surveys of the social mobilization-centered and state-centered studies, allowing readers to take stock of the rich literatures that have emerged over the past twenty to thirty years around LACEP. We help make sense of the extant research by highlighting the values, interests, policy preferences, and system incentives that shape the actions of social, economic, and state actors defining and contesting environmental policy in the region. Our organization of the sections reflects the focus of the LACEP literature on either the state or social actors. In our discussion, we also include the role of business and contend that the scarce attention paid to this actor is a shortcoming of the literature. We see this as a weakness of the society-centered approach, given that often business interests are independent from state interests in a majority of Latin America and the Caribbean. We believe there is a growing need to understand the environmental impact and especially the distinct preferred policies of different economic actors around the water–energy–food–environment nexus (Mahlknecht et al., 2020). This need will only become more urgent as climate policies dominate the political agenda of the region.

Second, this Element leverages the revealed preferences and incentives of social, economic, and state actors in the existing literatures to answer the central questions of distributive politics: *who* gets *what* under *which* conditions. We understand the *who* of this question to be either state, social, or economic actors seeking protection or status quo (i.e., the *what*) when facing a given environmental hazard under specific political and economic conditions. By asking the same questions of the social mobilization- and state-centered studies, this Element accomplishes its second goal, which is to help forge a new research agenda in LACEP focused on the distributive costs and benefits of degrading and protecting the environment. Furthermore, this agenda helps reveal when distributive benefits and costs are negotiated across crosscutting cleavages and when decisions that are adversarial to environmental protection are eventually overturned.

LACEP in Context

Before we begin our discussion of the LACEP literature, a brief comparison with other regions of the world is in order. Considering the region's impact on climate change and similar to other economies of the Global South, Latin America and the Caribbean countries have been low CO_2 emitters. However, size matters. Countries with larger populations and economies, such as Brazil, Mexico, Argentina, and to a lesser extent, Colombia and Venezuela, release significant amounts of methane gas and nitrous oxide. Within this group of countries, Brazil has the greatest impact on climate change, as the region's largest country and emitter of greenhouse gas (GHG) with the highest deforestation rates of the Amazon rainforest. Not surprisingly, Caribbean states contribute little if any to global warming but as island-nations are extremely vulnerable to climate change. Caribbean countries are at the frontline of exposure and vulnerability to sea-level rise and record-breaking storms, coupled with high levels of poverty. Islands like Puerto Rico, Cuba, Haiti, and the Dominican Republic – located in the so-called Hurricane Alley zone – are continually under threat of mass human casualties, infrastructure destruction, and losing population to climate-induced migration. In 2017 alone, the Caribbean nations suffered catastrophic devastation when they were hit in quick succession by Hurricanes Harvey, Maria, and Irma (Alcañiz and Sanchez-Rivera, forthcoming; Rodriguez-Díaz and Lewellen-Williams, 2020).

The Latin American and the Caribbean region exhibit some of the highest levels of income inequality in the world. Driven primarily by historic land concentration, poverty and land exclusion worsen even further for Indigenous and Afro-descendant citizens (Ahmed et al., 2021; Mollett, 2014; Pragier, 2019; Vélez et al., 2020; Vida, 2020; Wong, 2018; Zaremberg and Wong, 2018). In Latin America and the Caribbean, Indigenous and Afro-descendant citizens are less likely to have access to safe drinking water and sewage systems, and to hold formal rights to land and property (Ahmed et al., 2021; Freire et al., 2015; Zegarra et al., 2007). Similarly, and as the literature on environmental justice shows, impoverished ethnic and racial communities disproportionately overlap with industrial sites that pollute, pressure, and degrade their natural environment (Freire et al., 2015). Consequently, conflict stemming from water and land insecurity, and mining and other extractive developments intersect – and are exacerbated by – social inequalities resulting from ethnicity and race.

Finally, Latin American and Caribbean countries are extremely biodiverse. Brazil alone holds approximately 20 percent of global water supply and biological diversity (see the Convention on Biological Diversity). Conversely, due to climate change, global warming, and demographic and agricultural stressors, the region's biodiversity is at high risk. In 2018, Latin America and the Caribbean had 629 mammal species, 1,716 fish species, 1,117 bird species, and 5,439 plant species under threat (United Nations Environmental Program and the World Conservation Monitoring Centre, and International Union for Conservation of Nature, Red List of Threatened Species).

Figures 1 through 4 compare Latin America and the Caribbean to the other great regions of the world (Sub-Saharan Africa, South Asia, East Asia and the Pacific, Europe and Central Asia, and North America) along four key indicators of climate change. As the figures clearly show, the region's emission levels in CO_2 (Figure 1) and methane (Figure 2) are low and comparable to emissions from Sub-Saharan Africa and South Asia. However, when we examine renewable energy consumption (Figure 3), Latin America and the Caribbean appear closer to North America and Europe and Central Asia. In contrast, when we look at the region's forest area as a percentage of land area (Figure 4), Latin America and the Caribbean are ahead of all other regions, even with growing rates of deforestation.

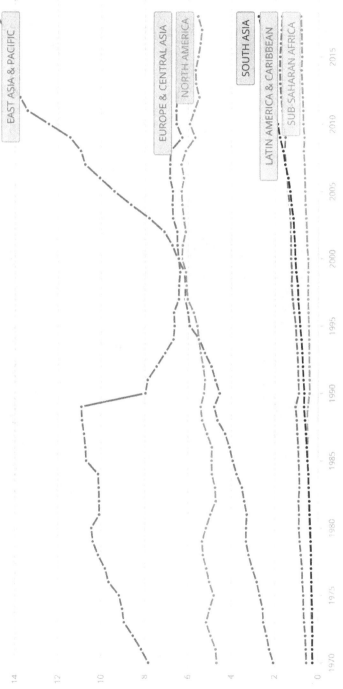

Figure 1 CO$_2$ emissions (kt) for Latin America and the Caribbean, Sub-Saharan Africa, South Asia, East Asia and the Pacific, Europe and Central Asia, and North America

Note: Data for up to 1990 are sourced from Carbon Dioxide Information Analysis Center, Environmental Sciences Division, Oak Ridge National Laboratory, Tennessee, United States. Data from 1990 are CAIT data: Climate Watch. 2020. GHG Emissions. Washington, DC: World Resources Institute. Available at: climatewatchdata.org/ghg-emissions.

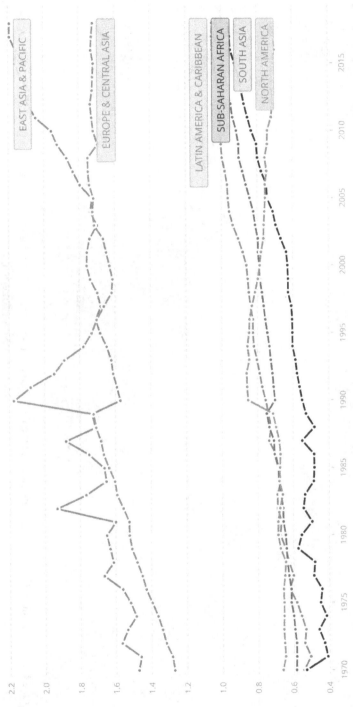

Figure 2 Methane emissions (kt of CO_2 equivalent) for Latin America and the Caribbean, Sub-Saharan Africa, South Asia, East Asia and the Pacific, Europe and Central Asia, and North America

Note: Data for up to 1990 are sourced from Carbon Dioxide Information Analysis Center, Environmental Sciences Division, Oak Ridge National Laboratory, Tennessee, United States. Data from 1990 are CAIT data: Climate Watch. 2020. GHG Emissions. Washington, DC: World Resources Institute. Available at: climatewatchdata.org/ghg-emissions.

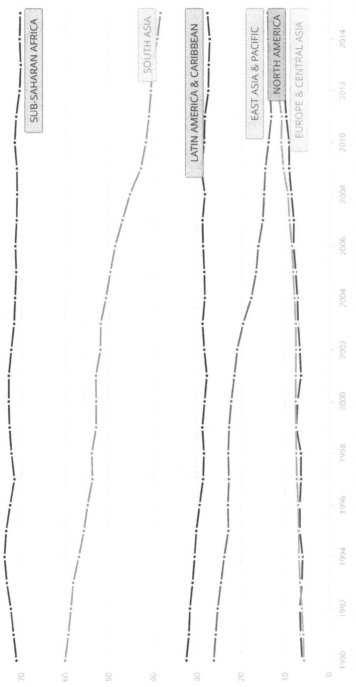

Figure 3 Renewable energy consumption (% of total final energy consumption) for Latin America and the Caribbean, Sub-Saharan Africa, South Asia, East Asia and the Pacific, Europe and Central Asia, and North America

Note: World Bank, Sustainable Energy for All (SE4ALL) database from the SE4ALL Global Tracking Framework led jointly by the World Bank, International Energy Agency, and the Energy Sector Management Assistance Program.

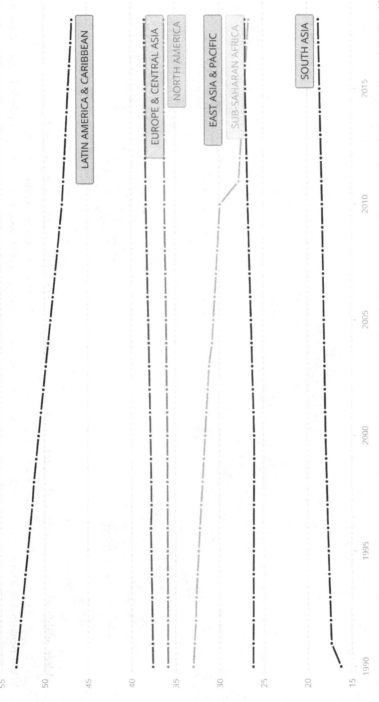

Figure 4 Forest area (% of land area) for Latin America and the Caribbean, Sub-Saharan Africa, South Asia, East Asia and the Pacific, Europe and Central Asia, and North America

Note: Food and Agriculture Organization, electronic files and web site.

2 The Social Mobilization Perspective in Latin America and the Caribbean

The Latin American and Caribbean Environmental Politics (LACEP) literature that centers on the mobilization of social groups vying for environmental protection mostly inquires about the conditions under which actors forge collective action. Environmentally oriented social mobilization refers to collective action that aims to influence decisions regarding the political and economic regulation of the interaction between human beings and their environment (Christel and Gutiérrez 2023). As depicted by the literature, a mosaic of actors, issues, preferences, and strategies are involved in environmental mobilization in Latin America and the Caribbean (Baver and Lynch, 2006a; Carruthers, 2008a; Christen et al., 1998; Lewis, 2016; Urkidi and Walter, 2011; Viola, 1992; Walter and Urkidi, 2016). Despite the difficulties of unifying the literature under one theoretical framework, here we identify key themes that help make sense of the complex and rich experiences under analysis.

In this section we review the social-mobilization-oriented LACEP literature by addressing three sets of questions. The first research questions are related to the basics of environmental mobilization: Who are the actors engaged? What types of actions do they perform? How do they interact (or not) among each other? At what scales does mobilization occur? How does the political and economic context affect environmental mobilization? While two basic types of groups are present in most accounts of environmental mobilization, elite organizations and grassroots groups, studies disagree with regards to the interaction among them and offer different narratives of the types of actions and the mobilization scales involved.

The second set of questions centers on the state: How do social actors relate to the state? Does social mobilization have an impact on environmental policies? The relationship of social organizations with the state is not explicitly examined by many studies on environmental mobilization. Still, while studies focused on grassroots organizations emphasize or assume an adversarial relation to the state, the studies that center on alliances, coalitions, and networks describe better the interaction between social (and sometimes economic) actors and states agencies. As we will see, this literature shows that the success or failure of environmental claims rely to a large extent on the collaboration between state and social actors.

Finally, we examine the full distributive nature of environmental mobilization as revealed by the literature. What are the preferences and interests of nonstate actors? How is social mobilization linked to distributive benefits and costs? Much of the literature tends to assume that environmental conflicts reinforce existing cleavages in society. That is, behind every environmental conflict there is a distributive inequality between those who benefit from the appropriation and contamination of natural resources and those who suffer from the socio-environmental costs of that appropriation. The winners of this inequitable distribution are, as a rule, the large companies and the state that carry out or promote large investment projects. The losers are typically the populations affected by the projects promoted by large companies and the state. This distributive conflict is somewhat taken for granted and not sufficiently examined empirically across much of the literature.

As the following pages show, however, we find that some studies do analyze in detail the different conflicting positions and even find evidence that every now and then some economic actors may side with the protectionist coalition. This LACEP scholarship reveals that crosscutting cleavages form around environmental conflict, as they do around other social and political battles. These studies also show that local populations can sometimes defeat the powerful and reverse the course of state decisions. This generally happens when (1) broad alliances are built, and within which diverse social actors coalesce with political and state actors that legitimize and drive their claims, and especially when (2) two other conditions are met: the economic activity in question is in the projected stage and it collides with entrenched local economies.

Environmental Mobilization

Who are the actors engaged in environmental mobilization? Even though the universe of Latin American and Caribbean environmental organizations is extremely heterogenous, a distinction between two basic types is common to most of the literature: elite organizations and grassroots groups (Baver and Lynch, 2006b; Bull and Aguilar-Støen, 2015; Christel and Gutiérrez, 2023; Christen et al., 1998; Foyer and Dumoulin Kevran, 2017; Gutiérrez and Isuani, 2014; Lewis, 2016). Yet studies disagree as to how the two interact. As we will see, some see them as disconnected social forces, while others emphasize the alliances, coalitions, and networks that are built between elite and grassroots organizations (Christel and Gutiérrez, 2023).

Elite organizations are private organizations officially recognized by the state. They tend to be professionalized organizations, relying on paid staff and with office installations. Critically, they may be funded by international donors or through state programs. On the other hand, grassroots groups are made up of individuals bound together by and for a common issue. Their work is often triggered by local challenges and conflicts. Their resource base relies on volunteer work and tends to be independent from international funding (Christel and Gutiérrez, 2023; Gutiérrez and Isuani, 2014).

Elite organizations have played an important role in the advocacy of biodiversity and conservation (Baver and Lynch, 2006a; Christen et al., 1998; Foyer and Dumoulin Kevran, 2017; Lewis, 2016; Viola, 1992). Grassroots groups may include Indigenous and peasant communities, local populations and environmentalists, or community organizations. The presence of grassroots groups is reported all across the region, from Chile and Argentina to Mexico and the Caribbean (Baver and Lynch, 2006a; Bebbington, 2011; Bebbington and Bury, 2013; Bull and Aguilar-Støen, 2015; Carruthers, 2008a; Christen et al., 1998; Foyer and Dumoulin Kevran, 2017; Gutiérrez and Isuani, 2014; Svampa, 2012; Walter and Martínez-Alier, 2010; Walter and Urkidi, 2016).

Environmental mobilization in Latin America and the Caribbean involves a great variety of issues and frames. Elite environmental organizations are usually seen as pursuing specific conservationist issues, many times under the ideational and financial influence of international environmental organizations. For instance, conservation, biodiversity and the protection of virgin lands have been central issues of elite environmental activism in Costa Rica, Venezuela, Ecuador, Brazil, and the Caribbean countries (Christen et al., 1998; Jácome, 2006; Lewis, 2016; Viola, 1992).

Regarding grassroots groups, studies link them to a predominance of territorially bounded issues that revolve around the defense of a lifestyle closely connected to local livelihood and traditional subsistence practices. Examples of grassroots groups grievances include claims around the impacts of tourism in Martinique (Burac, 2006), the struggles for land tenure in Costa Rica (Cordero Ulate, 2017), the defense of community development initiatives in Mexico (Foyer and Dumoulin Kevran, 2017), and the opposition to mining in many localities of the Andean and Central American countries (Bebbington, 2011; Bebbington and Bury, 2013; Cisneros, 2016; Urkidi and Walter, 2011; Walter and Urkidi, 2016).

The LACEP literature on social mobilization offers different framings of environmental issues, but we can make a general distinction between status quo and radical frames (Christel and Gutiérrez, 2023). On the one hand, elite environmentalists typically make their claims from a perspective that holds that environment sustainability is compatible with, or at least does not oppose directly, economic development. Their most recurrent framings reflect different approaches to sustainable development or other perspectives that propose environmental protection as a way to secure natural resources-based economic development (Aguilar-Støen and Hirsch, 2015; Jácome, 2006; Lewis, 2016). On the other hand, studies that focus on grassroots groups account for more critical or radical frames that question the capitalist economy and even the concept of sustainable development. Radical framing includes different environmental justice frames (Acselrad, 2004; Carruthers, 2008a; Lynch, 2006; Martínez-Alier et al., 2015; Urkidi and Walter, 2011), as well as other frames that dispute the bases of the current economic development model, such as popular ecologism and eco-resistance (Lewis, 2016; Martínez-Alier, 2007) or ecofeminism (Leguizamón, 2020; Sempertegui, 2021).

What types of actions do social actors perform? The dichotomy between elite organizations and grassroots groups is often accompanied by the distinction between institutional and contentious modes of action (Christel and Gutiérrez, 2017, 2023). From a dichotomous view, elite organizations or groups that are highly formalized mostly use institutional modes of action and usually engage in collaborative interactions with the state (Bull and Aguilar-Støen, 2015; Christen et al., 1998; Lewis, 2016). Instead, grassroots groups mostly perform contentious actions and tend to view the state in an adversarial way (Baver and Lynch, 2006a; Carruthers, 2008a; Martínez-Alier et al., 2016) or have a limited capacity (or interest) to influence state decisions (Bull and Aguilar-Støen, 2015; Edwards and Roberts, 2015).

Nevertheless, some studies claim that both grassroots and elite organizations can combine contentious and traditional or institutionalized modes of action (Bebbington et al., 2011; Christel and Gutiérrez, 2017, 2023; Domínguez, 2008; Lemos, 1998; Viola, 1992). Protests, marches, blockades, strikes, signed petitions, and the like combine with institutionalized channels. Examples of mixed tactics and strategies abound in Latin America and the Caribbean. These may include legal actions in subnational conflict against mining in Argentina (Christel, 2020); consultations and referendums over forest management in Costa Rica (Cordero Ulate, 2017) and large-scale mining projects across the region (Walter and Urkidi, 2016); and lobbying in antipollution struggles in Brazil (Lemos, 1998; Lemos and Looye, 2003) and Puerto Rico (McCaffrey and Baver, 2006).

As many studies show, the combination of institutionalized and contentious actions becomes more likely when different types of actors work together around common issues and goals. In her study of pollution control policies in the City of Cubatão, Brazil, Lemos holds that

> although the identification of shared goals is a necessary condition for the establishment of the alliance between progressive technocrats and popular movements, it does not mean that both groups have to agree on how to achieve them. Therefore, even if technocrats and popular movements have the same general goals, they will most likely adopt different pursuit tactics. Thus, popular movements tend to assume a confrontational and radical

position regarding their demands, while technocrats are more willing to compromise as a means to achieve their goals (Lemos 1998, p. 85).

How do social actors interact with each other? Part of the literature sees elite organizations and grassroots groups as disconnected (if not opposing) actors that pursue separate issues, resort to distinctive modes of actions (e.g., institutionalized versus contentious actions) and engage in different types of relationships with the state (Acselrad, 2010; Baver and Lynch, 2006a; Christen et al., 1998; Lewis, 2016). Through this dichotomous lens, grassroots activism tends to be seen as locally restricted and with no connection to other actors, while elite organizations are perceived to have stronger ties to international organizations and donors.

Instead of disconnected or opposing forces, as mentioned earlier in the text, some studies find that elite organizations and grassroots groups work together around common issues and goals, as part of the same alliance, coalition, or network (Alvarado Merino, 2008; Bebbington et al., 2011; Bratman, 2015; Cisneros, 2016; Cordero Ulate, 2017; Domínguez, 2008; Foyer and Dumoulin Kevran, 2017; Gutiérrez, 2018; Hochstetler and Keck, 2007; Lemos, 1998; Lemos and Looye, 2003; McCaffrey and Baver, 2006; Merlinsky, 2013b; Svampa, 2012; Valdés Pizzini, 2006; Velázquez López Velarde et al., 2018; Viola, 1992). These networks at times even include state actors, as we also discuss in the next section. This approach of forging broad coalitions can be found in multiple examples across Latin America and the Caribbean. They include the struggle against military control of the Vieques Island in Puerto Rico (McCaffrey and Baver, 2006); the defense of transhumance pastoralism and ancestral lands in Northern Argentina (Domínguez, 2008); the resistance to open-pit mining across Latin America (Cisneros, 2016; Urkidi and Walter, 2011; Walter and Urkidi, 2016); and the influence of environmental coalitions in Argentine environmental policymaking (Gutiérrez, 2018).

A few studies go a step further and acknowledge the participation of economic actors in pro-environment alliances, coalitions, and networks. Perhaps the most notorious case is the participation of local rural producers in the resistance to open-pit mining and other large-scale projects across the region (Alvarado Merino, 2008; Cartagena Cruz, 2017; Cisneros, 2016; Foyer and Dumoulin Kevran, 2017; Urkidi and Walter, 2011). Other local business sectors (such as tourism) have also been involved in such cases as the opposition to open-pit mining (Christel, 2020; Walter and Martínez-Alier, 2010) and pulp mill projects in Argentina (Alcañiz and Gutiérrez, 2009).

Existing research examines how wide and flexible frames, as well as reframing processes, favor the formation of alliances, coalitions, and networks. For example, the reframing of local grievances initially cast as economic ones to an environmental justice frame that focuses on public health, environment, and human rights has been crucial in the Vieques, Puerto Rico conflict (McCaffrey and Baver, 2006). Furthermore, the convergence of environmentalist and Indigenous groups goes hand in hand with the integration of sustainable development language and collective rights discourses in Northern Argentina (Domínguez, 2008). Finally, the gradual transition toward a sustainable development frame alongside the adoption of a socio-environmentalist perspective marked the consolidation of the Brazilian environmental movement (Hochstetler and Keck, 2007; Viola, 1992).

The LACEP scholarship that examines how different types of actors work together finds little consensus in the extent to which common issues and goals can be formulated by environmental networks. To some (perhaps most), environmental alliances, coalitions, and networks are engaged in single issues (such as mining, pollution, or deforestation) and pursue territorially bounded goals (Alvarado Merino, 2008; Bebbington et al., 2011; Bratman, 2015; Cisneros, 2016; Cordero Ulate, 2017; Domínguez, 2008; Lemos, 1998; Lemos and Looye, 2003; McCaffrey and Baver, 2006). Others see grassroots groups and elite organizations as members of a same social movement that transcends territorial conflicts and single issues and are concerned with multiple issues related either to the depletion of natural resources or to urban problems and the access to public goods and services (Cartagena Cruz, 2017; Foyer and Dumoulin Kevran, 2017; Merlinsky, 2013b; Svampa, 2012; Valdés Pizzini, 2006; Velázquez López Velarde et al., 2018; Viola, 1992).

At what scales does mobilization occur? Environmental mobilization occurs at different scales in Latin America and the Caribbean (Christel and Gutiérrez, 2023). As the LACEP literature shows, it may take place at the local, national, regional, or international level. Some studies also show that a same movement may scale up several levels.

Much of the scholarship see grassroots groups as place-based and locally restricted, like the case of mining conflicts in Guatemala (Hogenboom, 2015) and the many experiences of popular ecologism across the region (Martínez-Alier et al., 2016). On the other hand, elite organizations are typically perceived to work at different levels within a country and may even have foreign ties to international organizations and donors (Christen et al., 1998; Lewis, 2016). In all, these studies suggest that the probability of scaling up tend to be higher for elite organizations than for grassroots groups. Yet, others find evidence of how conflict can grow from the local to the state or provincial level, and to the national level, or even go global. Examples of this include the grassroots struggle over the Belo Monte dam in Brazil (Bratman, 2015) or the many mining disputes across the region, which may start at the district level and end in national conflict (Alvarado Merino, 2008; Walter and Urkidi, 2016).

Similarly, the social movement literature involves multiple-scale analyses as it centers on networking which may take place throughout the different levels within a given country and also at the international level (Bebbington et al., 2011; Svampa, 2012; Velázquez López Velarde et al., 2018; Viola, 1992). In many cases, international partners are extremely important because they provide not only political, economic, logistic, and knowledge support from abroad but also access to international arenas. National and transnational networks, alliances, and coalitions may help transcend the initial "particularism" of local conflicts as scholars show (Carey, 2009; Foyer and Dumoulin Kevran, 2017; Hochstetler and Keck, 2007; Svampa, 2012; Walter and Urkidi, 2016). Hochstetler and Keck hold that socio-environmentalists struggling to protect forests in the Amazon were joined by local and national state agents as well as international actors (Hochstetler and Keck, 2007). They maintain:

> Those resisting the dominant model developed alliances, information networks, and linkages to other parts of Brazil and the world. By accelerating the flow of information and resources through these networks, activists hoped gradually to enlarge the space available for endogenously generated development alternatives (Hochstetler and Keck, 2007, p. 142).

How do political and economic conditions affect environmental mobilization? The influence of political and economic conditions, under which environmental mobilization occurs, is a critical area of study and does not receive always the attention it deserves. Politico-economic factors are particularly important to scholars that see environmental mobilization through the lens of social movements as well as by studies on extractivism. A few studies explicitly trace the causal mechanisms linking politico-economic conditions and environmental mobilization. Exceptions include Bebbington's (2011) and Bebbington and Bury's (2013) edited volumes on mining and hydrocarbons in South America as well as McCaffrey and Baver's (2006) study on Puerto Rico.

As a general characterization, we can distinguish between studies that focus on national and international politico-economic conditions. For example, Velázquez López Velarde and his colleagues focus on the impact of Mexican domestic political factors, such as political opportunity structure, political alliances and coalitions, state capacities, and public opinion (Velázquez López Velarde et al., 2018). Christen and her coauthors, as well as Hochstetler and Keck discuss how the national political system affects environmental politics in several Latin American countries, confirming the importance of democracy and political openness for environmental mobilization (Christen et al., 1998; Hochstetler and Keck, 2007).

On the other hand, McCaffrey and Baver show how changes in the political opportunity structure at the international level affect the prospects of environmental mobilization in Puerto Rico (McCaffrey and Baver, 2006). Svampa as well as Bebbington and his coauthors pay more attention to the international economic context and hold that the renewed focus of domestic economies on primary products fuels the struggles over land and the environment (Bebbington et al., 2011; Svampa, 2012;). Bridging both levels of analysis, Viola refers to the international context and the domestic opportunity structure as a general but critical background of the Brazilian environmental movement (Viola, 1992).

The Link between Environmental Mobilization and the State

How do social actors relate to the state? Does social mobilization have an impact on environmental protection? Although the state dimension is not examined explicitly by many studies on environmental mobilization, we can distinguish in the literature two main ways in which social actors interact with the state: adversarial versus collaborative interaction (Bebbington et al., 2011; Christel and Gutiérrez, 2017; Christen et al., 1998; Gutiérrez and Isuani, 2014; Lewis, 2016; Martínez-Alier et al., 2016; Svampa, 2012; Valdés Pizzini, 2006).

On one side of the equation, some authors show that grassroots groups (Martínez-Alier et al., 2016) and environmental movements (Bebbington et al., 2011; Svampa, 2012; Valdés Pizzini, 2006) typically engage in an adversarial relation to the state. On the opposite side, professional or elite organizations are supposed to build a more collaborative interaction with the state (Christen et al., 1998; Gutiérrez and Isuani, 2014; Lewis, 2016). A clear example of this opposition can be found in the distinction advanced by Lewis between "eco-resisters" and "eco-dependents" within Ecuadorian environmentalism (Lewis, 2016). While "eco-resisters" (voluntary organizations and local groups) engage in adversarial relations to the state, "eco-dependents" (typical professional organizations) adopt a mostly cooperative approach to the state as they rely heavily on state and international funds. In a similar vein, Christen and her coauthors indicate that scientific expertise may afford professional organizations more channels to government access in such countries as Costa Rica, Mexico, Venezuela, and Brazil (Christen et al., 1998). Valdés Pizzini, on the other hand, argues that what unites the disparate groups that form the Puerto Rican environmental movement is their opposition to the state:

> What unites these types of groups into a movement is the fact that they are part of a process in which civil society challenges the state in an

'environmental field' consisting of the processes and problems that affect the social, cultural, and biotic health of the community, alter ecosystems, threaten species, and change traditional culture and resource use patterns. This field is characterized by the exclusion of local communities and organizations from environmental policy and decision making (Váldez Pizzini, 2006, p. 46).

A more balanced view is rendered by studies that point to different cases of alliances and networks in which local or municipal governments and other state agents join or support grassroots groups, professional organizations, and other social actors (Cartagena Cruz, 2017; Christel, 2020; Cordero Ulate, 2017; Gutiérrez, 2018; Hochstetler and Keck, 2007; Lemos, 1998; McCaffrey and Baver, 2006; Walter and Urkidi, 2016). These works represent a rare attempt to overcome the somewhat artificial state–society divide. For example, local governments may be crucial allies in the opposition to the central government.

As Walter and Urkidi show in their study of mining conflict in Peru, Argentina, Guatemala, Ecuador, and Colombia, local authorities bring formal legitimacy to the demands of an environmental coalition, especially through the implementation of bottom-up consultations and plebiscites (Walter and Urkidi, 2016). In the same vein, Viola contests the state–society divide that informs much of the literature by arguing that the Brazilian environmental movement grew across all sectors in Brazil: civil society, the state, business, and science (Viola, 1992). Viola also argues that Brazilian environmental organizations and groups underwent a progressive process of professionalization during the 1980s, with an ensuing shift toward a more moderate relation to the state (Viola, 1992).

Also in Brazil, Lemos argues:

> Progressive technocrats within Cetesb [São Paulo state environmental agency] and popular movements in Cubatão formed a temporary alliance that provided both groups with sufficient political clout to push for pollution control implementation (Lemos, 1998, p. 76).

Lemos also reveals a type of division of labor that can be found in other cases: while bureaucrats provide expertise and authority, popular organizations supply mobilization and public attention (Lemos, 1998).

As stated in the introduction of this Element, state–society collaboration is possible due to the heterogeneity of state actors and policy preferences. Studies show divisions across the environmental claims and positions of state actors. They also show that this variation is crucial to understanding the political effects of environmental mobilization, as the success or failure of environmental claims rely to a large extent on the ability to find allies within the state and the collaboration between state and social actors (Castro et al., 2016; Christen et al., 1998; Gutiérrez, 2020, 2018, 2017; Hochstetler and Keck, 2007; Lemos, 1998; Lemos and Looye, 2003; McCaffrey and Baver, 2006; Walter and Urkidi, 2016). As Silva argues, when social mobilization aims to introduce changes in environmental governance:

> [the] potential for change involves engagement with established political systems at the international, national or subnational scale or combinations of them to effect institutional and policy reforms (Silva 2016, p. 331).

In a similar vein, Gutiérrez (2017) shows how the alliance between social organizations, environmental state agencies, and individual legislators helped offset the embeddedness of forestry and rural interests within the state and allowed for the passing and the implementation of a national forest protection law in Argentina.

The Distributive Politics of Environmental Mobilization

What are the actors' preferences and interests? How is social mobilization linked to distributive benefits and costs? Implicitly or explicitly, most of the literature on environmental mobilization assumes the existence of a general distributive conflict between social actors, on the one hand, and large economic actors and/or state actors, on the other. Social actors are not only opposed to certain economic activities, but also to some state projects (such as conservation projects and a number of infrastructure projects) that they consider affect their interests and their livelihood (Barandiaran and Rubiano-Galvis, 2019; Gerlak et al., 2020; Hochstetler, 2020).

Often, these distributive conflicts are assumed and not sufficiently examined empirically in much of the literature, which, as we have seen so far, focuses fundamentally on the characteristics of mobilization. However, as we stated earlier in the text, some studies do analyze in detail the different conflicting positions, while a few others even find that economic actors may be on the side of the protectionist coalition occasionally. In these final passages of the mobilization section, we focus on these studies because they are the ones that most clearly identify the distributive nature of environmental politics in Latin America and the Caribbean.

The environmental justice and environmentalism of the poor scholarship reveals the distributive dimension of environmental mobilization. This literature defines environmental justice as the distribution of environmental goods and ills between those who benefit from the appropriation and contamination of natural resources and those who suffer from the socio-environmental costs of that appropriation (Acselrad, 2006, 2008, 2010; Alimonda, 2006, 2008; Carruthers, 2008a; Martínez-Alier, 2004; Martínez-Alier et al., 2010; Urkidi and Walter, 2011; Walter and Urkidi, 2016). The winners of this inequitable distribution are, as a rule, the large companies and the state that carry out or promote large investment projects. The losers are typically the people directly affected by the projects promoted by large companies and the state. While the former highlight the positive economic and social impacts of their initiatives, the latter deny these benefits and affirm that both their health and their livelihoods and cultural traditions are threatened by them (Alvarado Merino, 2008; Merlinsky, 2013c; Paredes, 2006; Paredes and Kaulard, 2020; Urkidi and Walter, 2011).

From this perspective, the power relations between winners and losers are essentially asymmetric, since the latter are systematically excluded from decision-making (Carruthers, 2008a; Paredes and Kaulard, 2020). For this reason, some authors highlight that the claim for environmental justice is often accompanied by a claim for procedural justice to the extent that the mobilized actors also demand changes in the governance mechanisms and greater participation in the decisions that affect them (Carruthers, 2008b; Urkidi and Walter, 2011; Walter and Martínez-Alier, 2010; Walter and Urkidi, 2016). Carruthers describes two types of inequality as follows:

> In addition to *distributional inequity*, one of the pillars of environmental justice is a concern about *procedural inequity*, propelling a quest for greater political participation and more authentic citizenship. Environmental grievances gain legitimacy when representatives of affected communities demonstrate not just disproportionate exposure but deliberate exclusion from the political decisions that determine the locations and the risk levels of environmental threats (Carruthers 2008b, p. 8).

Given existing asymmetries of power and the exclusion from decision-making, some studies seem to reaffirm again and again that the affected populations cannot change the course of policy decisions and are always losers (Acselrad, 2010, 2008, 2006; Alimonda, 2008, 2006; Baver and Lynch, 2006b; Bratman, 2015; Carruthers, 2008a; Paredes and Kaulard, 2020). However, in their study of opposition to open-pit mining, Walter and her coauthors carry out detailed analyses showing under what conditions affected locals can become winners (Urkidi and Walter, 2011; Walter and Martínez-Alier, 2010; Walter and Urkidi, 2016).

For example, Walter and Martínez-Alier (2010) examine the local opposition to a gold mining project in the small city of Esquel, in the Argentine Patagonian Andes. They show how a group of neighbors managed to form a new grassroots organization (Asociación Vecinos Autoconvocados or Self-Convened Neighbors Association) and how two coalitions quickly emerged in opposition.

The pro-mining coalition – made up of provincial government officials, the local chamber of commerce, and the union of construction workers – argued that mining would be a beneficial activity for the economic development of the region and that it could be environmentally sustainable if technologies and adequate controls were in place. The anti-mining coalition – forged by the Self-Convened Neighbors Association, the local water cooperative, small businesses, Mapuche organizations, and regional environmental organizations – countered that mining could not be environmentally sustainable and was incompatible with their vision of local development. They insisted that economic development efforts should focus on existing nonpolluting activities, such as agriculture, forestry, and tourism. The municipal government had a central role in this conflict. Initially, it supported the pro-mining coalition, but in the face of increasing social mobilization against mining, it changed its position and approved a referendum in order to decide whether or not to accept the gold mining project. At the same time, the courts accepted an appeal for protection (*recurso de amparo*) that suspended the approval of the project until new impact assessments were carried out. The referendum vote rejected the mining project and the municipal government passed an ordinance prohibiting the use of cyanide in mining operations. Even though it had the authority to disregard both the outcome of the referendum and the municipal ordinance, the provincial government canceled the Esquel gold mining project and the provincial legislature passed a law prohibiting open-pit mining.

Urquidi and Walter (2011) examine the Pascua Lama mining case in Chile, which shares many similarities with the Argentine case but resulted in a different outcome. Both in Esquel and Pascua Lama, two coalitions were formed in favor and against mining. In both cases, the anti-mining coalition argued that mining would exacerbate the unequal distribution of economic goods and environmental ills and was incompatible with the dominant economies of the region (agriculture in Pascua Lama, agriculture and tourism in Esquel). It also called for more equitable decision-making that would recognize the right of local populations to determine which economic activities should be promoted in their territories. According to the authors there is one key difference between the Argentine and Chilean cases. Indigenous organizations in Pascua Lama did not join the anti-mining coalition (even though they were opposed to the project) and the local government did not support the latter either. This would explain why the anti-mining coalition did not have the same strength in Pascua Lama as it did in Esquel.

Walter and Urkidi (2016) reach similar conclusions in a large study of sixty-eight community consultations on metal mining across Peru, Ecuador, Argentina, Guatemala, and Colombia. Community consultations are promoted by grassroots organizations and other local actors who oppose mining. In all cases, the actors promoting the consultation combine two major claims: (1) mining endangers their livelihoods and (2) local communities have the right to participate in decisions that affect their livelihood and to define how these decisions be made. The design of community consultations stems from local organizations themselves, varies from one locality to another, and may differ from the free, prior, and informed consent consultation for Indigenous communities established by Convention 169 of the ILO. Those who defend mining (the mining industry and the central government) typically reject, ignore, or criminalize community consultations. Some local and provincial governments, instead, legitimize them and transform them into "formal" institutions through local ordinances. Local government support is key if grievances against mining are to reach regional, national, and even international levels and have any real chance of shutting down the mining projects, as they observe in the Peruvian and Argentine cases.

Walter and Urquidi stress that community consultations do not seek to "bypass government" but to anchor their legitimacy in the local state:

> We could say that the legitimacy of consultations is, in part, both a cause and a consequence of the hybrid alliances formed between local governments and social movements. The involvement of local governments and the diverse positions adopted within state and government bodies regarding community consultations reflect the heterogeneity of interests and values across these structures. This feature of consultations points to the need to further problematize the role of governments and the state in environmental governance frameworks. Hybrid institutions led by civil society, such as community consultations, do not necessarily aim to 'bypass governments' (as pointed out by Delmas and Young, 2009) but, on the contrary, to anchor part of its legitimacy in some of its bodies (local governments) (Walter and Urquidi 2016, p. 317).

Taken together, these studies show that competing values and interests regarding economic development and the environment come into play in distributive conflicts. They also show that local populations can sometimes defeat the powerful and reverse the course of state decisions around the utilization of natural resources. This generally happens when broad alliances are built within which diverse social actors (e.g., Indigenous and peasant communities, local, national, and/or transnational environmental organizations), local producers, and small businesses network with political and state actors. Broad coalitions help to legitimize the environmental claims of local stakeholders.

A number of scholars highlight the importance of broad coalitions and show that some local economic actors can also be part of these alliances (e.g.,

Alvarado Merino, 2008; Cartagena Cruz, 2017; Christel, 2020; Foyer and Dumoulin Kevran, 2017; Hochstetler and Keck, 2007; McCaffrey and Baver, 2006; Paredes, 2006). These studies show that environmental distributive conflicts do not have a single likely outcome whereby industry and governments are always winners and locals are always losers. Indeed, studies tend to find that local populations win when three conditions are met: (1) the economic activity in question is still in the projected stage or is not fully operational; (2) the proposed activity collides with entrenched local economies and the mobilization against it is articulated around the resulting opportunity cost; and (3) social actors manage to build a broad alliance with the participation of political and state actors who are key to reversing any prior decisions.

The probability that crosscutting coalitions will overturn state and economic decisions they oppose increases when the targeted activity is in its initial phase, given that it is not embedded yet in local or domestic interests (Bratman, 2015; Christel and Gutiérrez, 2021). However, success become more difficult when the opposed activity is already at work because opponents have to face embedded economic, social, and political interests that make more difficult the building of a broad or effective coalition. Similarly, when a proposed activity collides with entrenched local economies, broad coalitions are more likely to form and succeed because social and economic actors may find a common enemy. This makes it more likely that opponents can secure the support of relevant political actors who are sensitive to the electoral mood and economic interests of their constituency.

The merging of social and economic interests against a given activity may strengthen crosscutting coalitions in two different ways: either through the direct incorporation of economic actors into the coalition, as in the examples of Esquel, Vieques, and other cases mentioned above, or through the packaging of socio-environmental and economic concerns made by social organizations, as illustrated by the opposition to open-pit mining in Mendoza, Argentina (Christel, 2020). All in all, these studies point to the importance of paying more attention to the forging of crosscutting coalitions and the conditions under which they form and work in order to unpack and further our understanding of the distributive dynamics of environmental politics.

3 The State Perspective in Latin America and the Caribbean

The extensive state-focused literature on LACEP examines the role that the public sector plays in environmental protection. This scholarship reveals that the state is not a uniform or monolith actor nor does it always provide environmental protection (Alcañiz and Gutiérrez, 2020a; Amengual, 2016; Bauer,

2015; Hochstetler and Keck, 2007; Ponce and McClintock, 2014; Urkidi and Walter, 2011). State actors often behave in contradictory ways, sometimes enforcing environmental regulations and at other times, violating those same policies (Alcañiz and Gutiérrez, 2020a; Amengual, 2016; Armesto et al., 2001; Azócar et al., 2005; Bauer, 2015; Carruthers, 2001; Fernández-Milmanda and Garay, 2019; Hochstetler and Keck, 2007; Ponce and McClintock, 2014; Urkidi and Walter, 2011). The state-focused literature also reveals how officials engage with social and economic actors and how different policy preferences and values shape the process of environmental protection (Abers and Keck, 2019; Armesto et al., 2001; Bauer, 2015; Carruthers and Rodriguez, 2009; Gonzalez, 2021; Hochstetler, 2020; Orta-Martínez et al., 2018; Paredes, 2018; Paredes and Kaulard, 2020; Valladares and Boelens, 2017). Finally, a consistent yet not always explicit theme in this literature is that distributive costs and benefits are revealed once the state becomes involved in the contest over environmental protection (Armesto et al., 2001; Azócar et al., 2005; Carruthers and Rodriguez, 2009; Gonzalez, 2021; Orta-Martínez et al., 2018; Paredes, 2018; Paredes and Kaulard, 2020; Valladares and Boelens, 2017).

In this section, we review the state-focused literature by answering key questions that inquire into the different ways in which state actors engage in LACEP and provide or withdraw environmental protection. As the state is a complex entity, we begin by offering some analytical clarity on the different ways in which it manifests. We then structure the rest of the section around key questions that the LACEP literature helps answer. Understanding that state-sanctioned protection of the natural world may present in varied manners, we ask what types of actions the state performs to protect the environment. We also inquire into the different state actors who engaged in environmental protection. Next, we ask how state actors interact with each other. Latin American and Caribbean governments in recent decades are just as likely to support green policy as they are to pursue actions that overturn it. Consequently, we examine the impact of social and economic actors, such as activists and economic producers, when they coalesce together or against the state. How do social and economic actors relate to the state? Critically, we are interested in the distributive effects of state–society interaction in LACEP. Thus, the final question we ask in this section is: How do state actions in environmental policy determine distributive costs and benefits?

As we discuss in the introduction of this Element, we caution the reader that the boundaries between the state-focused and society-focused scholarship are somewhat artificial. The state – in its multiple manifestations – appears when-ever there is conflict over the environment and natural resources. Students of environmental and climate politics, even those who locate the onus of political change elsewhere (on the organizational strength of social movements or the economic power of business, for example), are forced to contend with the sanctioning power of the government. Whether the government protects the status quo, pushes for reform, sides with victims of environmental suffering, or backs the offender, all of these policy alternatives entail state regulation. Having stated our caveat, we define state-focused studies as those that purposefully situate the political fight over the natural world within the government or identify it as a key player in the struggle.

Environmental Protection

What types of actions does the state perform to protect the environment? If the goal of environmental policy is to protect the natural world, any discussion of it must make explicit the different ways in which the state delivers this protection. Environmental protection by the state may manifest in different ways. The state may protect the environment proactively through regulation, administrative decisions, legislation, and judicial decisions (Aguilar-Støen, 2018; Alcañiz and Gutiérrez, 2020a; Amengual, 2016; Bauer, 2015; Fernández-Milmanda and Garay, 2019; Herrera and Mayka, 2020; Hochstetler and Keck, 2007; Kauffman and Martin, 2017; McAllister, 2008; Ponce and McClintock, 2014; Urkidi and Walter, 2011). State intervention may range from upholding existing laws, pushing for reform, or adjudicating between stakeholders when there is conflict. With this in mind, environmental policies can be understood as the authoritative decisions made by public officials in the different branches of government at the national, subnational, or international level with the goal of protecting the natural world and the human communities that reside in it.

Environmental protection is achieved through the authoritative decisions that regulate the use of natural resources with the objective of conserving or preserving them (Bull and Aguilar-Støen, 2015). As discussed in the introduction of this Element, protection may entail preventing the occurrence of a probable harm or correcting the effects of an existing one. While the role of social mobilization is critical in explaining policy reform and governance schemes, ultimately it is the state that institutionalizes protection of the natural world by fiat (Alcañiz, 2016; Durazo, 2004; Guevara Sanginés, 2005; Gutiérrez, 2018; Herrera and Mayka, 2020; Kauffman and Martin, 2017; Micheli, 2002; Narváez, 2007; Ponce and McClintock, 2014).

Conversely, through actions and inactions, the state may also institutionalize environmental harms (Fernández-Milmanda and Garay, 2019; Guevara Sanginés, 2005; Herrera, 2017; Herrera and Mayka, 2020; Kauffman and Martin, 2017; Paredes and Kaulard, 2020; Silva et al., 2002). This is especially clear in Latin America and the Caribbean, where in recent decades the state has created a record number of programs to protect the environment while simultaneously pushing the agricultural, energy, and mining frontier in order to receive windfall profits (Alcañiz and Gutiérrez, 2020a, 2020b; Fernández-Milmanda and Garay, 2019; García-López and Arizpe, 2010; Kauffman and Martin, 2017; Kohl and Farthing, 2012; Paredes and Kaulard, 2020). Indeed, some of the scholarship that examines the environmental costs of economic development even overlaps with the study of the rentier state, which derives its rents from the exploitation of natural resources (Arce, 2014; Arce et al., 2018; García-López and Arizpe, 2010; Gustafsson, 2017; Kohl and Farthing, 2012; Ponce and McClintock, 2014; Valladares and Boelens, 2017).

Who are the state actors engaged in environmental protection? In conceptualizing the state, one initial distinction is whether researchers focus on the national, international, or subnational level. To begin with, a wealth of social science scholarship studies national state agencies and institutions in LACEP (Abers and Keck, 2013; Aguilar-Støen, 2018; Alcañiz, 2016; Bauer, 2015; Challenger et al., 2018; Ebeling and Yasué, 2009; Hochstetler and Keck, 2007; Hochstetler and Viola, 2012; Mumme, 2007; Paredes and Kaulard, 2020; Wismer and Lopez de Alba Gomez, 2011).

For example, Hochstetler and Keck identify the national ministry of the environment, the preferences of its authorities, and federal environmental laws in Brazil as key actors engaged in the protection of the natural world and the institutional arena within which this occurs (Hochstetler and Keck, 2007). Similarly, in *Practical Authority*, Abers and Keck study the evolution of public committees and agencies in charge of permits for water use in Brazil (Abers and Keck, 2013). They find that this less rigid conceptualization captures more accurately how the federal state is many things at the same time (e.g., party and judge, actor and arena) and anticipates the different ways in which state intervention in water policy can create distributive costs. Students of LACEP get a realistic description of the Brazilian federal state and the distributive effects of government action are made explicit. The authors describe the federal state in Brazil as:

> Like other organizations, governments mobilize organizational capabilities and implement projects. They coordinate employees, interact in networks, hold and participate in debates, elaborate plans, generate and use scientific knowledge, occupy offices and buildings, provide services, build roads and hospitals. To get done, these activities do not require a threat of violent coercion or the legitimacy of a state that monopolizes the legitimate use of that threat (Abers and Keck 2013, p. 8).

The subnational state is another critical area of study of LACEP (Alcañiz and Gutiérrez, 2020a, 2020b; Amengual, 2016; Bauer, 2015; Challenger et al., 2018; Fernández-Milmanda and Garay, 2019; Gagnon-Légaré and Prestre, 2014; Gustafsson, 2017; Pacheco-Vega, 2019; Paredes and Kaulard, 2020; Ponce and McClintock, 2014; Wismer and Lopez de Alba Gomez, 2011). The importance of this level of analysis stems from the fact that environmental problems vary greatly within a country, regardless of its wealth, political system, and regime. Geography may also directly affect the type of environmental problem. Water pollution can only happen where there are bodies of water (albeit, often in areas more densely populated), deforestation only occurs where there are forests, loss of biodiversity happens in the natural habitats of flora and fauna. Variation in local state attributes and resources help explain differences in environmental policy outcomes. Critically, variation across types of subnational states helps explain differences in distributive gains and losses stemming from environmental conflict at the local level.

Both the national and subnational LACEP literatures draw attention to the ties forged between state and stakeholders and find evidence that the resulting networks serve as supplemental sources of regulation resources for impoverished bureaucracies (Alcañiz, 2016; Amengual, 2016; Gagnon-Légaré and Prestre, 2014; Paredes, 2018). The Latin American and Caribbean environmental politics research reveals that state officials draw on their connections to key societal and economic groups, rarely engaging in environmental protection in isolation. Studies show that these networks are especially common in subnational environmental agencies (Abers and Keck, 2013; Alcañiz, 2016; Amengual, 2016; Gagnon-Légaré and Prestre, 2014; Gustafsson, 2017; Hochstetler and Keck, 2007; Paredes, 2018).

The study of networks between state and nonstate actors is of critical importance to help develop a distributive research agenda in LACEP. Indeed, who the state picks as a partner in environmental protection often foretells who benefits from the distributive consequences of state action. In her study of prior consultation laws, extractive industries projects, and Indigenous communities in Peru, Paredes highlights the role of the Ombuds Offices (Defensorías del Pueblo) and their ties to human rights groups (Paredes, 2018). She finds that this network helped increase the representation of marginalized communities, especially Indigenous ones, in the Ombuds Office's push for environmental protection:

> The Ombuds Office's ability to recruit talented staff from human rights NGOs and university networks contributes to its identity as part of a larger 'community' with clear normative commitments ... the Ombuds Office remains an exception [to Peru's reputation as a weak state] due to its excellence in defending human rights. The result is a deep engagement with civil society;

a territorial reach with twenty-eight functioning local branches; the effective representation of marginalized groups; and rapid answers to the call of citizens when human rights are reported in danger (Paredes 2018, p. 104).

How do state actors interact with each other? Much of the state-focused LACEP scholarship involves more than one government actor, scaling up, down, or sideways in their negotiation of environmental protection (Alcañiz, 2016; Alcañiz and Gutiérrez, 2020a; Fernández-Milmanda and Garay, 2019, 2020; Gagnon-Légaré and Prestre, 2014; Mumme, 2007; Paredes and Kaulard, 2020). State actors at different levels of government may even compete with one another. Without doubt, scarce resource and competing jurisdictions may lead to intra-state conflict (Alcañiz, 2016; Alcañiz and Berardo, 2016; Alcañiz and Gutiérrez, 2020a; De Pourcq et al., 2017; Gizelis and Wooden, 2010; Pacheco-Vega, 2020). However, different state units are just as likely to cooperate on environmental and climate policies within and across different agencies and jurisdictions.

At the most local level of environmental policy, we find evidence of scaling up and direct engagement with other levels of the state. For example, in Herrera's (2017) study of water policy in the outskirts of Mexico City, she finds that meaningful reform is effective in part when upper-levels of the state support it. That is, this occurs when the municipal government passes local regulation and the subnational (and national) state reinforces it. Similarly, Pacheco-Vega illustrates how local water policy in Mexico is connected to other levels of government and how municipalities are constrained both by local and federal politics (Pacheco-Vega, 2020).

The global nature of most environmental crises creates strong incentives for state actors – national and subnational – to coordinate internationally (Alcañiz, 2016; Alcañiz and Berardo, 2016; Mumme, 2007; Sears and Pinedo-Vasquez, 2011; Silva et al., 2002; Steinberg et al., 2001; Tecklin et al., 2011; Tigre, 2019; Torres Ramirez, 2019). The rise in the number of intergovernmental agreements on climate change, water and air pollution, and biodiversity loss in recent years attests to this. Increasingly, state units at all levels – such as bureaucratic agencies or congressional delegations – exchange information and know-how in order to address some of the shared urgent problems. International actors like policy experts, bureaucrats, and transnational activists also make up crosscutting coalitions with local groups. Intergovernmental coordination often entails distributive effects at the national level that may go unrecognized. The influence of powerful international actors on domestic environmental policy also helps decide the winners and losers of the benefits and costs of protecting the environment and of the deteriorating of natural resources. These domestic costs and benefits should be made more explicit by internationally focused LACEP state scholars.

Governments sometimes pursue international strategies in order to escape intra-state conflict, which in turn enables the influence of international actors on domestic environmental policy either as lobbying actors or members of cross-cutting coalitions (Alcañiz, 2016; Alcañiz and Berardo, 2016; Sears and Pinedo-Vasquez, 2011; Silva et al., 2002; Tecklin et al., 2011; Tigre, 2019). For example, Alcañiz and Berardo (2016) show how regulatory agencies create intergovernmental governance consortiums to deal with common natural resources, like the five La Plata basin countries. In the late 1960s, Argentina, Bolivia, Brazil, Paraguay, and Uruguay created the Intergovernmental Coordinating Committee of the Countries of the La Plata Basin and signed the Treaty of the La Plata Basin (Alcañiz and Berardo, 2016). National regulatory agencies dominate transgovernmental cooperation in water comanagement, due to the high concentration of hydroelectric plants in the basin.

Similarly, Bolivia, Brazil, Colombia, Ecuador, Guyana, Peru, Suriname, and Venezuela created the Amazon Cooperation Treaty Organization (ACTO) in 1998 to address some of the shared problems of the world's largest rainforest, such as water insecurity, biodiversity loss, and deforestation through an intergovernmental mechanism. Tigre analyzes the role of ACTO in addressing common crises and supporting the Nationally Determined Contributions (NDCs) under the United Nations Framework Convention on Climate Change (UNFCCC), the commitments in emission reductions offered by all Framework participating countries (Tigre, 2019). As a strategy for local state actors attempting to solve domestic environmental conflict, scaling up sometimes falls short, especially if there is not full cooperation and coordination among parties (Alcañiz, 2016; Tigre, 2019).

Powerful international agents engage with state actors at different levels of government to influence domestic environmental politics. One interesting example of this is the United Nations REDD+ (Reducing Emissions from Deforestation and forest Degradation) program, a Payments for Environmental Services (PES) scheme. This program (REDD+) seeks to limit deforestation by paying landowners not to cut down trees and has a number of pilot programs in the Amazonian states of Brazil (Börner and Wunder, 2008). The program is complex in that it has multiple different sources of funding which find their way to numerous recipients. Donors include corporations like Petrobras, Ford and Packard; international organizations, like the World Bank and the United Nations Environmental Program; individual countries, like Norway and the United States; and Brazilian state governments (REDDX Tracking Forest Finance Project). Recipients include the federal government of Brazil, subnational governments, and the private and nonprofit sectors (REDDX Tracking Forest Finance Project). As deforestation is one of the main ways in which Brazil and other large countries in the region contribute to climate change, international organizations and OECD states strongly endorse these market-friendly forest PES schemes (Zhouri, 2010). The distinct interests of the many international stakeholders involved in the program increase local communities' resistance to this program and remind us of the distributive dimension of climate policy tools (Alcañiz and Gutiérrez, 2020a; Zhouri, 2010).

In their analysis of cross-national reforms of forest legislation, Silva and his coauthors discuss how international actors participated in Bolivia's forest reform (Silva et al., 2002). They find that the United States Agency for International Development (USAID) and the United Nations Food and Agriculture Organization (FAO) succeeded in influencing the design of the Bolivian forest reform (Silva et al., 2002). Similarly, Tecklin and his coauthors explain the evolution of Chile's environmental regime as "driven primarily by external forces linked to economic globalization" (Tecklin et al., 2011, p. 880). In their account, international actors helped favor market-friendly environmental protection, nudging the Chilean government when it appeared less committed to laissez-faire policies. Chile, the first South American country to adopt a market-centered economy in the 1970s under the Pinochet dictatorship, was particularly receptive to a deregulation approach to deal with its many environmental problems. "Heading into the reform, the Chilean government was split as to whether to push for a full environmental ministry or some lower profile organizational form ... A final inducement toward a minimalist approach to reform was provided by the World Bank which offered a US$17 million loan for the reform process" (Tecklin et al., 2011, p. 887). Steinberg shows an even deeper influence of international actors in their analysis of transnational training of local actors in the early days of the environmental movement in the region (Steinberg, 2001).

Sears and Pinedo-Vasquez (2011) offer another example of how international and national state actors engage with one another and shape the adoption of domestic forest policy. Their research helps explain the persistence in Peru of an informal patronage system of logging concessions (*habilitaciones* in Spanish), where timber industrialists and political patrons function almost as an illegitimate alternative state or para-state. To great extent, unlawful, clientelistic-based logging endure in the Peruvian Amazon region because local states have weak capacity and are adversely "shaped by development aid conditionality, international conservation agendas and trade agreements" according to the authors (Sears and Pinedo-Vasquez, 2011, p. 624). In their account, international-level incentives interact with the preferences and institutional limitations of the domestic political system, where a weak state lacks the technical capacity and political will to enforce forest laws.

Alcañiz (2016) also examines international system-level incentives and finds that government experts who staff the national environmental agencies of Latin American and Caribbean countries seek international technical cooperation when their budgets and resources are cut. Environmental bureaucrats work with peer programs, accessing new know-how and technology to help supplement existing state capacity. Similar to the bilateral activists that Steinberg finds in Bolivia and Costa Rica, Alcañiz describes environmental government scientists who scale up and "go international" to overcome the lack of investment in skills by national and subnational states (Alcañiz, 2016; Steinberg, 2001). Many of these transnational policy experts end up participating in domestic cross-cutting coalitions.

Global negotiations around the UNFCCC offer an opportunity to examine how Latin American and Caribbean states use the international dimension strategically to resolve internal conflict (Edwards and Roberts, 2015; Torres Ramirez, 2019; Viola and Franchini, 2014). Brazil, the region's giant and the greatest contributor to climate change, shows the depth of its political polarization when its government navigates international climate policy (Hochstetler and Viola, 2012; Viola and Franchini, 2014; Viola and Franchini, 2017). The struggle between the (ultra) conservative and reformist sectors of the Brazilian political system, which appears intensely in all policy areas, is front and center in the country's negotiation of climate policy (Viola and Franchini, 2014; Viola and Franchini, 2017).

The Mexican state delegation, as another example, led by the Environment and Natural Resources Secretariat (SEMARNAT) and the National Institute of Ecology and Climate Change (INECC), pursued dual alliances as a Latin American country and a neighbor and key ally of the United States. On the one hand, Mexico pushed for common goals with other developing countries in the region (e.g., the Pacific Alliance with Chile, Colombia, and Peru). On the other, Mexico advocated for many of the US-desired objectives, such as greater commitment by developing countries to mitigation efforts and seeking to include different types of natural gas in the mitigation inventory (Torres Ramirez, 2019).

The Link between State and Society

Similar to our discussion in Section 1 on how social mobilization and state responses are connected, we ask in this subsection how social and economic actors relate to the state. The state-focused LACEP literature tells us that regardless of the goals, strategies, and crosscutting coalitions forged by different actors, environmental protection is ultimately decided within the state (Aguilar-Støen, 2018; Amengual, 2016; Bauer, 2015; Fernández-Milmanda and Garay, 2020; Gutiérrez, 2018; Herrera and Mayka, 2020; Hochstetler and Keck, 2007; Ponce and McClintock, 2014; Urkidi and Walter, 2011). Social and economic actors, such as activists, business associations, and other stakeholders, appeal to state officials to secure their support or opposition to environmental protection. Thus, the question of state access – who receives government protection and who is left out – is critical (Aguilar-Støen, 2018; García-López and Arizpe, 2010; Herrera and Mayka, 2020).

State access helps determine how the costs and benefits of environmental protection often correlate with the costs and benefits of economic development. As environmental justice scholars have pointed out repeatedly, it is often the same groups that on the one hand, benefit and on the other, lose out from economic development and protection (Carruthers, 2008a; da Rocha et al., 2018; Martínez-Alier, 2008; Perez Guartambel, 2006; Rodríguez et al., 2015; Sundberg, 2008; Ulloa, 2017; Urkidi and Walter, 2011). To put it differently, the reason environmental policy entails direct distributive consequences is because state certification of policy entails access and protection for some, and exclusion for others. We posit in this Element that the question of state access should be made explicit in the study of LACEP, as it reveals the distributive effects of environmental politics. In the following pages, we examine some of the ways in which society and state connect in order to change or maintain environmental policy and how the LACEP literature covers them.

We can identify three groups of economic and social actors that seek to influence and access the state on environmental protection in Latin America and the Caribbean: business (Carruthers, 2001; Eakin and Lemos, 2006; Ebeling and Yasué, 2009; Fernández-Milmanda and Garay, 2020, 2019; García-López and Arizpe, 2010; Gutiérrez and Jones, 2004; Hochstetler, 2020; Hochstetler and Viola, 2012), activists (Abers and Keck, 2009; Aguilar-Støen, 2018; Arce, 2014; Carruthers and Rodriguez, 2009; Ceddia et al., 2015; da Rocha et al., 2018; García-López and Arizpe, 2010; Herrera, 2017; Herrera and Mayka, 2020; Klepek, 2012; Kohl and Farthing, 2012; Riofrancos, 2020; Silva et al., 2018; Urkidi and Walter, 2011; Valladares and Boelens, 2017; Velásquez Runk, 2012), and to a lesser extent, experts (Aguilar-Støen, 2018; Alcañiz, 2016; Challenger et al., 2018; Gutiérrez, 2010; Kauffman and Martin, 2017). As we state in the introduction of this Element, activists and businesses are not monolithic actors nor can their policy preferences always be predicted (MacLeod and Park, 2011; Murphy and Bendell, 1997).

To begin with economic actors, the LACEP state literature focuses more on broader market forces and its effects on the body politic, and less on how businesses may organize politically to press for their preferred policy outcomes. Much of this scholarship shows that states with strong political and institutional mandates to privatize and deregulate enable the outsized influence of the private sector. In his work on environmental politics in Chile, Carruthers describes the legacy of neoliberalism of the Pinochet dictatorship (1973–1990) on environmental policy in the country (Carruthers, 2001). Chile is a paradigmatic case regarding state and business relations in environmental policy, and not surprisingly, it has received considerable attention from the LACEP literature. Much of the scholarship finds the relationships between state officials and economic actors to be highly symbiotic in the environmental sector:

> Such market solutions factor importantly into Chilean environmental policy. Consistent with Pinochet's vision, the state role is to be minimised. In several policy areas the focus has been on property rights regimes to 'get the prices right' and avert the 'tragedy of the commons'. To their advocates, environmental protection can best be achieved through privatisation, export promotion and the maximisation of economic growth. . . . In a deregulated political climate, meaningful enforcement power is almost non-existent—a far cry from successful European-style ecological modernisation. This lack of state capacity and autonomy is in fact a defining feature of the Chilean business climate, prized by developers and business interests unaccustomed to constraint (Carruthers, 2001, p. 349).

Alcañiz and Gutiérrez (2020a) examine the impact of economic forces and the state on deforestation in Argentina. They show how market pricing of farmland and institutional separation of economic and green missions in subnational environmental ministries help explain differences in deforestation rates of native forests across the country's provinces (Alcañiz and Gutiérrez, 2020a). They find that when local governments have greater financial capacity, measured in the amount of forest PES they disburse, and a sole green mission (i.e., bureaucratic specialization), deforestation declines (Alcañiz and Gutiérrez, 2020a). But high farmland prices, as an indicator of agricultural producers' strength, fuels deforestation at a much greater rate than subnational PES reduces it (Alcañiz and Gutiérrez, 2020a).

More directly, Fernández-Milmanda and Garay (2020) examine the power of agricultural producers in order to explain the implementation of federal forest protections by subnational states. The authors seek to explain why the effective enforcement of these protections vary among a small number of northeastern provinces in Argentina. Their argument is based on a distributive logic: in active agricultural frontiers driven by the national government, governors are especially conflict-averse and will weaken either the design or implementation of anti-deforestation programs depending on the political strength of large agricultural producers *vis a vis* conservationists (Fernández-Milmanda and Garay, 2020; Hochstetler, 2021).

In demanding protection and reform, social actors frequently engage with the state through protest (Aguilar-Støen, 2018; Arce, 2014; Arce et al., 2018; Carruthers and Rodriguez, 2009; García-López and Arizpe, 2010; Orta-Martínez et al., 2018; Pellegrini and Arsel, 2018). But the role of activists and social organizations in shaping environmental protection continues to grow beyond contentious politics. State officials increasingly rely on community organizations for environmental policy design, adoption, and implementation (Alcañiz and Gutiérrez, 2009; Amengual, 2016; Herrera, 2017; Nicolle and Leroy, 2017; Torres Ramirez, 2019). State actors draw on community organizations for know-how and technical expertise, as well as to shore up government efforts in policy monitoring and enforcement (Abers and Keck, 2009; Alcañiz and Gutiérrez, 2009; Amengual, 2016; Bragagnolo et al., 2017; Carruthers and Rodriguez, 2009; Nicolle and Leroy, 2017; Torres Ramirez, 2019). This is especially so when states exhibit weaknesses in their enforcement capacity, which may create incentives for bureaucrats to follow the lead of societal actors in order to facilitate the implementation of new policies (Amengual, 2016; Amengual and Dargent, 2020; Fernández-Milmanda and Garay, 2020).

Informal ties and networks between social actors and public officials are critical entry points for activists to influence environmental policy, as discussed above (Abers and Keck, 2013; Amengual, 2016; Carruthers and Rodriguez, 2009; Gudynas, 2009; Hochstetler and Keck, 2007; Nicolle and Leroy, 2017; Paredes, 2018). Informal ties between environmental activists and like-minded bureaucrats often forge permanent networks within which policy know-how and proposals are exchanged. Social actors, both elite and grassroots organizations, influence bureaucrats and legislators this way (and vice versa). Furthermore, these networks serve as recruiting grounds for state appointments (Abers and Keck, 2013; Hochstetler and Keck, 2007; Paredes, 2018).

In their study of how Brazil adopted protected areas in Northern Amazonia, Nicolle and Leroy describe the connection between state and social actors in similar terms, as a loose coalition of government officials and activists (Nicolle and Leroy, 2017). They find that these advocacy coalitions form around the goal of fighting deforestation in the Amazon region, albeit with different strategies, which in turn are negotiated inside the coalition (Nicolle and Leroy, 2017). Amengual (2016) explains state access at the subnational level as a result of networks that connect state, social, and economic actors. This type of state–society networks also provides political access to foreign climate policy. For example, the Mexican government attends the annual Conference of the Parties of the UNFCCC meeting with a coalition of civil society actors that represent the business sector and environmental advocacy groups (Torres Ramirez, 2019).

At times, networks forged by state, social, and economic actors become institutionalized. Paredes' case study of the Peruvian Ombuds Office discussed above, shows how an autonomous state agency can offer activists the opportunity to influence environmental regulation at the subnational level (Paredes, 2018). The 2007 constitutional assembly of Ecuador offers another example. This temporary institution of representative democracy became a focal point for Ecuadorian activists to push for the Rights of Nature (RoN) to be enshrined in the country's new constitution. Environmental activists sought to influence the process in different ways, but especially by activating existing ties to political actors. A few activists were elected themselves, many became staff to constitutional representatives, and yet others partnered up with international NGOs to have an enduring impact on the political system (Gudynas, 2009).

The Brazilian Forum on Climate Change, created by former president Michel Temer (2016–2018) to work on the country's commitments under the Paris Agreement, is another good example of this (Hochstetler, 2021). The group has achieved some concrete goals, such as drafting climate policy (Hochstetler, 2021). It also attempts to forge strategies for subnational states to work around the anti-climate policies of the federal government and because of that has become a political target of President Jair Bolsonaro (2019–present).

Courts are another key arena within which societal grievances meet the state (Alcañiz and Gutiérrez, 2009; Herrera and Mayka, 2020; Kauffman and Martin, 2017; McAllister, 2008; Urkidi, 2010; Whittemore, 2011). The number of lawsuits filed to uphold environmental protection by crosscutting coalitions of elite and grassroots social organizations, environmental experts, and even state bureaucrats has grown exponentially in recent decades. In Ecuador, activists have taken legal action and sued the state to uphold the Rights of Nature (RoN) (Kauffman and Martin, 2017). In Brazil, the role of prosecutors in enforcing existing environmental laws continues to grow as well, especially in the Amazonian states. Already in 2001, the Amazon region was a focal point of litigation: "The caseload of federal prosecutors in [the Amazonian state of] Pará is indicative of the priority placed on environmental prosecution: in 2001, over half of civil cases and about one-third of criminal cases concerned environmental harm" (McAllister, 2008, p. 5).

In her book on Environmental Protection and Legal Institutions in Brazil, McAllister surveys the legal changes occurring across the region:

> Like Brazil, many Latin American countries have written new constitutions that guarantee the right to a healthy environment. Colombia and Paraguay have also constitutionally charged their Ministerio Público with defending environmental interests. Ombudsman offices in Argentina (Defensoría del Pueblo) and Costa Rica (Defensoría de los Habitantes) have powers to protect citizens rights, including environmental rights, against unlawful governmental acts (McAllister 2008, p. 7–8).

Finally, scientists and experts also forge strong ties to state bureaucrats and participate in crosscutting coalitions with the aim of overturning adverse state decisions (Aguilar-Støen, 2018; Alcañiz, 2016; Challenger et al., 2018; Gutiérrez, 2010; Kauffman and Martin, 2017; Pielke Jr., 2007; Steinberg, 2001). In fact, many expert bureaucrats who staff the environmental ministries are scientists themselves and maintain close ties with their colleagues outside the agency and the larger bureaucracy (Alcañiz, 2016). Often, scientists access the state through institutional partnerships with public universities and research centers. These partnerships result from formal and informal networks of state officers and scientists, and can be domestic and international.

A transnational example of a partnership among state, expert, and agribusiness in the area of training is the Costa Rican Tropical Agricultural Research and Higher Education Center (CATIE in Spanish), with expertise in the sustainable management of natural resources in the region. Other domestic examples include the recruitment pipeline that may exist between public universities and the research units of state agencies with an environmental mission (Alcañiz, 2016).

The Distributive Politics of State Environmental Protection

How do state actions in environmental policy determine distributive costs and benefits? As our examination of social and economic groups in the LACEP literature reveals, the state does not always protect the environment. Frequently, state actors are slow to react to risks and hazards and seek to maintain the status quo. More critically, state officials sometimes help create those same risks and hazards. This is especially true in contemporary LACEP. During the 2000s global commodities boom – the super cycle of high commodity prices that benefited Latin American fuels and food exports – governments in the region sought to profit from it. Paradoxically, the same governments that adopted some of the most stringent environmental protections to date were the same helping further active frontiers in agricultural production and mining, which in turn created more environmental degradation and conflict (Alcañiz and Gutiérrez, 2020a; Arce, 2014; Fernández-Milmanda and Garay, 2020; Kohl and Farthing, 2012; Ponce and McClintock, 2014; Valladares and Boelens, 2017; Zimmerer, 2011).

Examples abound of Latin American leaders supporting new restrictive environmental legislation while at the same time contributing to the causes that led to the need for state protection. Under President Cristina Kirchner, the Argentine state signed multiple trade and investment agreements with China on soy – which effectively kept expanding the country's agricultural frontier – while at the same time adopting the most restrictive forest and glaciers laws (Alcañiz and Gutiérrez, 2020a). In 2005, the president of Brazil, Lula da Silva, and the country's minister of environment, Marina Silva, created a "mosaic" of 32,000 square mile protected areas across the Amazon rainforest to try to stop the expansion of the soy-fueled agricultural frontier responsible for extreme deforestation rates (Branfors and Borges, 2019). The Lula administration even helped put in place moratoriums on soy and beef produced in the Amazon region. Yet at the same time, through generous funding by the National Development Bank, Brazil expanded industrial and energy development throughout the country (Hochstetler, 2020).

As discussed throughout this Element, crosscutting coalitions – whereby distinct actors, including representatives of the business and advocacy sectors, state officials and policy experts come together to uphold or pass environmental protection – often become winning coalitions (Alonso et al., 2008; Alvarado Merino, 2008; Cartagena Cruz, 2017; Christel, 2020; Foyer and Dumoulin Kevran, 2017; Gutiérrez, 2018; Hochstetler and Keck, 2007; Madariaga and Allain, 2020; McCaffrey and Baver, 2006; Paredes, 2006). These broad coalitions reveal the distributive nature of environmental protection as it groups distinct –sometimes even opposing – actors by political costs and benefits. This is similar to our discussion in Section 1 on how social mobilization and state responses are bridged by these broad alliances.

The legislature is a key institutional arena within which costs and benefits stemming from environmental policy are negotiated and distributed. Tecklin and his coauthors explain the central role the national congress in Chile has in the political process, enacting laws regulating fisheries, mining, and the historic 1994 National Environmental Framework Law (Tecklin et al., 2011). In Brazil, Castro Pereira and Viola examine recent setbacks in environmental regulation and squarely locate the distributive fight around environmental policy in the Brazilian national congress (Pereira and Viola, 2019). The authors draw up a list of new laws that reversed environmental protections, starting in 1996 and increasing in the 2000s during the commodity boom, when the largest number of regulations were overturned. This reversal trend has been especially exacerbated under the presidential administrations of Michel Temer and Jair Bolsonaro (Gerlak et al., 2020). Recent anti-environmental legislation includes the weakening of state safeguards of water systems (i.e., "granting licensing exemption for water supply and sanitation projects") and the endangerment of the extremely vulnerable Indigenous communities of the Amazon rainforest (i.e., restricting the National Indian Foundation [Funai] powers) (Pereira and Viola, 2019, p. 97).

Judicial courts are increasingly adjudicating environmental conflict between state, social, and economic actors, helping to determine who wins and who loses when the state takes action (Alcañiz and Gutiérrez, 2009; Armesto et al., 2001; Bauer, 2015; Herrera and Mayka, 2020; Kauffman and Martin, 2017; McAllister, 2008; Urkidi, 2010; Whittemore, 2011). Judicial sentences, consequently, have become a central mechanism of how the costs and benefits of environmental protection are distributed. In their analysis of Chile's market-friendly environmental regime, Tecklin and his coauthors show the courts helping to determine how developers use natural resources (Tecklin et al., 2011). They find that rather than being fully autonomous political agents, Chilean judges draw from the overall normative and political regime to decide frequently in favor of business. The distributive mechanism entailed in the courts' verdicts is clear to the authors: "In the absence of substantive environmental regulation, property rights have constituted the fundamental issue in most decisions" (Tecklin et al., 2011, p. 889).

Leite and Venâncio focus on Brazil's highest nonconstitutional court, the Superior Court of Justice or Superior Tribunal de Justiça (STJ), which determines jurisdiction and jurisprudence for both states and federal courts (Leite and Venâncio, 2017). They find that in recent years, the STJ has helped "standardize" environmental law in the country and uphold key principles, such as the right of nature, the precautionary principle, and sustainable development (Leite and Venâncio, 2017). Kauffman and Martin examine the courts in Ecuador, after the Andean country adopted in 2008 the world's first constitution that enshrined the right of nature – *Sumak Kawsay* in Quechua and *Buen Vivir* in Spanish – as a constitutional guarantee (Kauffman and Martin, 2017). Chapter 7, article 71 of the Ecuadorian constitution states the Rights of Nature (RoN): "Nature or *Pachamama*, where life is reproduced and exists, has the right to exist, persist, maintain and regenerate its vital cycles, structure, functions and its processes in evolution" (Rights of Nature Articles in Ecuador's Constitution). The distributive effects of environmental protection in Ecuador are revealed throughout: In the political process that led to the 2008 constitution including the RoN, in the decisions to contest or defend the RoN in court, in the pressure felt by state officials to uphold them, and in the backdrop of a record-breaking commodity boom in an oil-exporting county. These studies reveal the rising influence of judges as powerful arbitrators of LACEP.

Given our examination of the LACEP literature so far, we have identified two areas where we believe more research on the distributive politics of environmental protection is needed. First, the connection between social protests and state action is an area of study with great potential. How do the demands stemming from political mobilization trigger state involvement in environmental politics and policymaking? As discussed above, recent scholarship has begun fleshing out the precise ways in which these connections occur (Aguilar-Støen, 2018; Alcañiz and Gutiérrez, 2020a; Amengual, 2016; Bauer, 2015; Fernández-Milmanda and Garay, 2020; Gutiérrez, 2018; Herrera and Mayka, 2020; Hochstetler and Keck, 2007; Kauffman and Martin, 2017; McAllister, 2008; Ponce and McClintock, 2014; Urkidi and Walter, 2011). But often the protest spark that ignites an urgent state response is assumed rather than included in the causal mechanisms advanced by state-focused scholars.

We understand that this omission results in part from an inward-looking lens, whereby scholars of the public sector in LACEP focus overwhelmingly on the state. However, the disconnect between protest and policy makes the state-focused literature less sharp and effective. Furthermore, it misses out on the growing number of broad, crosscutting coalitions in LACEP that become winning coalitions, precisely because these alliances connect protestors with like-minded state experts and bureaucrats, and sometimes even business actors with aligned goals (Alonso et al., 2008; Alvarado Merino, 2008; Cartagena Cruz, 2017; Christel, 2020; Foyer and Dumoulin Kevran, 2017; Hochstetler and Keck, 2007; Madariaga and Allain, 2020; McCaffrey and Baver, 2006; Paredes, 2006).

A second omission of the state-centered literature is the role of business in environmental politics. While recent work is focusing more on business-state relations, the LACEP literature tends to overlook this key actor (Ebeling and Yasué, 2009; García-López and Arizpe, 2010; Hochstetler, 2020; Hochstetler and Kostka, 2015). If environmental politics are ultimately distributive, scholars should include economic preferences and interests in their analyses. Often, however, when LACEP state-centered studies include economic dimensions, they mostly reflect monolithic interests derived from structural conditions. But business actors hold distinct preferences in environmental protection depending on the economic sector to which they belong. For example, wind energy entrepreneurs in Brazil (and likely elsewhere) will prefer state intervention and subsidy whereas other energy business will oppose it (Hochstetler,

2020; Hochstetler and Kostka, 2015). In order to reveal the true distributive power of state decisions in LACEP, we must first understand the full agency of business and identify its varying preferences across sectors. If we seek to answer "who gets access to state protection?" and "who pays the price of that protection?" we cannot afford to leave hidden the policy preferences of all of the actors involved.

Furthermore, the centrality of business as an actor in LACEP can be clearly seen in state decisions in infrastructure and energy (Gerlak et al., 2020; Hochstetler, 2020). Undoubtedly, these are crucial areas of climate politics. Yet, with few exceptions, the literature on infrastructure, regulation, and the state in Latin America and the Caribbean has not fully caught up with the costs and impacts of these policy decisions for climate change (Bratman, 2015; Gerlak et al., 2020; Hochstetler, 2020). An extreme example of infrastructure investments pursued by state–business partnerships with enormous impact on the environment are hydroelectric mega dams. Touted as sources of renewable energy and low carbon-footprint, they also displace and destroy local human and ecological communities. While these costly infrastructure projects are overwhelmingly subsidized by governments, the rationale for their existence and the legal regime under which they are constructed and administered, typically result from state–business partnerships.

The Belo Monte Dam on the Xingu River in the state of Pará, Brazil is second in size only to the Brazilian-Paraguayan Itaipú Dam in the region. Plans for the construction of the Below Monte Dam started in the 1970s under the country's military dictatorship and quickly encountered great resistance from local Indigenous communities. A crosscutting coalition was forged early in the 1970s before construction in full began, including Indigenous and other community actors, clergy, transnational activists, and small-scale farmers (Bratman, 2015). In the face of this and ongoing opposition, the military state forged ahead and ensuing democratic governments and their business partners ensured operations almost four decades later despite proven accusations of corruption and socio-environmental losses (Bratman, 2015). What is more, the dam is seen as contributing significantly to climate change both by the deforestation required for its construction and the methane emissions resulting from its operation, which are found to be higher than initially calculated (Fearnside, 2017). Additionally, desertification driven by deforestation is creating droughts and historical water shortages in rivers like the Xingu, making the dam underperform in its generation of hydroelectricity (Fearnside, 2017).

Given our goal of uncovering the distributive costs and benefits of LACEP, it becomes clear to us that inequality will be a powerful determining force. That is, existing levels of social inequality act as critical sorting mechanisms, determining who gets which benefits from the state. Next, in the final section of this Element, we call for a new research agenda for LACEP aimed at explaining the political distribution of environmental protection.

4 A New Distributive Research Agenda for Latin American and Caribbean Environmental Politics

In this Element, we offer a state of the art of the study of LACEP. We examine how social demands for the use, conservation, remediation, and protection of natural resources are connected to decision-makers and other key political actors. We also examine how environmental social mobilization is shaped by the broader political and economic climate. We analyze how green elite organizations and grassroots groups differ in their goals and strategies as well as how social and economic actors access the state through contention or collaboration. In the two previous sections, we survey the critical questions of the LACEP scholarship that help reveal who become winners and losers in the fight over environmental policy.

Over the past two decades, the LACEP literature has gained depth and breadth on the different ways in which society and state engage in environmental politics. New and exciting research examines critical problems in the region. Recent LACEP scholarship includes in-depth studies on such diverse issues as water rights (Abers and Keck, 2013; Pacheco-Vega, 2019, 2020), deforestation politics (Fernández-Milmanda and Garay, 2019, 2020) and its intersection with gender (Alcañiz and Gutiérrez, 2020b), indigenous conflict in mining (Paredes, 2018) and in land use (Carruthers and Rodriguez, 2009), renewable energy (Hochstetler, 2020), constitutional reform (Kauffman and Martin, 2017), and environmental justice (da Rocha et al., 2018; Inoue and Moreira, 2017; Leguizamón 2020; Orta-Martínez et al., 2018; Ulloa, 2017; Urkidi and Walter, 2011). Furthermore, the LACEP literature has experienced a methodological transformation as well, going from mostly single-country case studies to larger comparative empirical studies (Alcañiz and Berardo, 2016; Arce et al., 2018; Ceddia et al., 2015; García-López and Arizpe, 2010).

As we discuss at length throughout this Element, at times the social mobilization and state literatures offer somewhat partial views of grievances and political demands by social and economic actors and the policy responses by state officials. While we still see a gap between the two literatures, we also recognize this seeming division of labor is becoming less defined. Nevertheless, the connection between society and state does not appear consistently in the

LACEP scholarship, and consequently, the distributive costs and benefits entailed in environmental protection are also obscured. In this Element, we call for this connection to be made explicit and distributive politics to become an organizing paradigm of LACEP research.

Thus, a central message of this Element is that environmental politics in Latin America and the Caribbean is as much about the protection of the natural world and its resources as it is about how environmental benefits and ills are distributed across society. In our survey of the LACEP scholarship we find that these distributive outcomes are contingent on how state, economic, and social agents engage with one another. We find hopeful evidence that affected communities – even historically excluded ones due to poverty and ethnic and racial disparities – can build winning coalitions. These coalitions, when forged by diverse actors representing different sectors of the polity, may even overturn state decisions that have the support of powerful economic actors, as we saw happen in the cases of the opposition to gold mining in Esquel (Argentina), the fight for pollution control in Cubatão (Brazil), and community consultations over mining across the region.

Environmental protection, as both government good and service, has become part and parcel of distributive politics. Similar to how governments reallocate income among constituents by delivering goods such as health care or public education, the preservation of the natural environment creates distributive effects of hazards, harms, and remediation. To explain the winners and losers of redistributive politics, the extant literature draws attention to the role of income (Calvo and Murillo, 2004; Szwarcberg, 2015), ideology (Dixit and Londregan, 1996; Kasara and Suryanarayan, 2015; Stokes, 2005) and parties and partisan networks (Calvo and Murillo, 2019; Carlin et al., 2015; Luna, 2010; Oliveros, 2018).

We see parallels in the LACEP literature. As the previous sections show, the LACEP scholarship examines how social organizations representing marginalized and lower income groups and economic actors press the government to rule in favor of their interests. Left of center politicians at times enact environmental regulations, even when often aggressively opposing the green demands of their own constituents (Edwards and Roberts, 2015; Gonzalez, 2021; Riofrancos, 2020). Lastly, the LACEP literature brings attention to how network ties and winning coalitions among advocates, lobbyists, and state ultimately help shape the distributive politics of environmental protection.

What are the distributive costs and benefits of environmental protection? As we discuss in the previous sections, they stem from regulations and the management of natural resources. Critically, they are determined by the interests and values of stakeholders. Some of the benefits of environmental regulation include sustainable land, less pollution, greater biodiversity, safe and obtainable water, stronger public health, slowdown of global warming, sustainable industry, and green jobs to mention a few. Preferences for these green goods and ecosystem benefits may also vary according to stakeholders' interests.

In the absence of environmental protection, climate change and environmental degradation continue to worsen, with grim consequences for the involved human and natural communities. The costs of environmental degradation can be measured by how it threatens the lives and livelihoods of affected populations. Environmental protection may also entail costs. These mostly stem from the implementation of new regulations on how industries do business. For example, mandating that mining companies regularly test groundwater quality in their area incurs in testing costs for the firm as well as potential fines if the regulation is violated. Our review of the LACEP literature revealed how economic actors – agricultural producers, mining companies, etc. – organize around perceived and real costs, often forging narrow alliances with social and state actors (Fernández-Milmanda and Garay, 2019; Gutiérrez and Jones, 2004; Hochstetler and Viola, 2012).

How costs and benefits linked to environmental regulation and economic activity distribute across communities is a central question of the environmental justice literature (Carruthers, 2008a; da Rocha et al., 2018; Leguizamón, 2020; Martínez-Alier, 2008; Perez Guartambel, 2006; Rodríguez et al., 2015; Sundberg, 2008; Ulloa, 2017; Urkidi and Walter, 2011; Zegarra et al., 2007). This distribution, according to the EJ scholarship, is not random. Rather, social inequality determines winners and losers of environmental protection. In Latin America as in the rest of the world, the subjects of the EJ literature are impoverished and marginalized communities. Disparities in gender, class, race, ethnicity, and citizenship status, to name a few, are the main drivers of unjust outcomes in the distribution of environmental costs and benefits. Environmental justice scholars predict distributive conflicts to fall along these lines; on one side, poor, Indigenous, and other excluded communities and on the other, developers, energy corporations, and often the state (Barkin and Lemus, 2016). Research in environmental justice tells us that it is the same individuals and communities who repeatedly miss out on economic benefits and receive little protection from environmental policy and little legal recourse to contest it.

We started this Element by asking the central questions of distributive politics from the LACEP literature: who profits from the appropriation of natural resources, who pays the costs of climate change and environmental degradation, and who benefits from state protections? We agree unequivocally with the EJ literature that existing patterns of inequality help answer these questions. Often the answer to these three questions are the powerful, the poor, and the powerful again, respectively. As the world's most unequal region of the world, all Latin American politics ultimately are affected by the politics of inequality. Especially environmental politics, as land concentration is a historic mechanism of economic inequality in the region. This in turn has a direct effect on environmental politics, which frequently stem from disputes over land access and use. The region's disparities will only deepen with the worsening of the global climate crisis and the brutal economic effects of the COVID-19 pandemic. As the EJ literature would predict, under these dire circumstances, environmental protection will entail even greater transfers of wealth. As a result, environmental politics has significant distributive consequences.

But our survey of the LACEP literature has also shown that within the constraints of social inequities there is room to maneuver in building winning coalitions. Actors respond differently to system-level politico-economic incentives. Even under rising party polarization and extremism, the scholarship in LACEP reveals that environmental politics contains crosscutting cleavages. As discussed in the previous sections, social actors coalesce into broad alliances with business and state actors and activists forge ties with like-minded bureaucrats and policy experts contributing to the design and implementation of environmental policy (Madariaga and Allain, 2020). Overlapping jurisdictions and multilevel governance help forge even broader coalitions where different visions and valuations of development and protection are negotiated. Latin American and Caribbean environmental politics, we find, are nuanced and networked and cannot be reduced to a single feud between society and firms. We find evidence that broader, crosscutting coalitions often explain how affected communities build winning coalitions and are able to revert state decisions – even when these decisions have vested interests attached to them (Madariaga and Allain, 2020; Urkidi and Walter, 2011; Walter and Martínez-Alier, 2010; Walter and Urkidi, 2016).

A key contribution of this Element is that it builds on an extensive literature that seeks to connect state and society and examine crosscutting coalitions. We believe the focus on a single actor may not be enough to explain policy changes or to understand the distributive politics of LACEP.

Our review of the LACEP literature reveals too how the fight over the costs and benefits of environmental protection has resulted in the strengthening of political agency of key actors in the region. Ethnic, racial, and gender identities increasingly shape political organization across the Americas. Communities of color and women have been excluded historically from the profits stemming from economic development and saddled with the evils of environmental degradation. But a survey of recent environmental conflicts also provides evidence of how marginalized communities have organized and become formidable political actors (Bebbington et al., 2011; Bose, 2017; Lewis, 2016; Paredes and Kaulard, 2020; Wong, 2018).

Unmistakably, Latin America has witnessed the rise of politically powerful indigenous movements in recent years. Their grievances and political organization have fueled a wave of protests that at times managed to paralyze entire governments in Ecuador, Chile, and Peru. Native peoples are central actors of the distributive politics of environmental policy, as their communities hold historic and present claims over land and other natural resources. Indigenous communities often hold formal stewardship over these resources, like community-based forests rights, and as a result are in the line of fire of both developers and state actors. Consequently, much of the activism and scholarship on environmental mobilization in Latin America, centers on indigenous politics, examining the grievances, interests, and preferences of native communities (Armesto et al., 2001; Azócar et al., 2005; Carruthers and Rodriguez, 2009; Gonzalez, 2021; Orta-Martínez et al., 2018; Paredes, 2018; Paredes and Kaulard, 2020; Valladares and Boelens, 2017). Increasingly, new research examines the heterogeneity of interests and preferences of environmental indigenous politics, reveling how crosscutting cleavages involve native and nonnative Latin Americans and how the former are not a monolithic group (Wong, 2018; Zaremberg and Wong, 2018; Pragier, 2019). We see this line of research as a key future agenda of environmental distributive politics.

Black and Afro-Latin Americans are also key actors of the distributive politics of environmental protection. Increasingly, the LACEP scholarship is uncovering the environmental plight of Afro-descendants, who traditionally have been erased from the history and political science books on the region (Lasso, 2019). New research examines the disproportionate effect of distinct land uses on Black communities in Brazil (Bledsoe, 2019; Futemma et al., 2015; Perry, 2009), Colombia (Ahmed et al., 2021; Vélez et al., 2020), Nicaragua (Vida, 2020), and Honduras (Mollett, 2014). Recent work also examines the role of race in disaster politics in Puerto Rico (Alcañiz and Sanchez-Rivera, 2021; Rodriguez-Díaz and Lewellen-Williams, 2020; Tormos-Aponte, 2018).

Scholarship focused on intersectional identities, especially of gender, ethnicity, and race, highlight the increased political role women play in environmental protection (Bolados García and Sánchez Cuevas, 2017). Indigenous women are at the forefront of conflict over land use and communal rights (Bose, 2017) as well as food production and eco-agriculture (Zuluaga-Sánchez and Arango-Vargas, 2013). Women are also key beneficiaries of state investments in protection, such as in programs for land PES (Alcañiz and Gutiérrez, 2021).

Finally, as social concerns about climate change are notably increasing in the region, we identify two phenomena that have not yet been sufficiently studied because of their novelty. Inspired by Greta Thunberg and the Fridays for Future campaigns, youth activism against climate change has spread across the region demanding national and international authorities and leaders the implementation of urgent solutions. Second, the climate emergency and a surge of record-breaking natural disasters increase and worsen environmental displacement. The desperate plight of climate migrants in the region – many internally displaced in their home countries – is increasingly calling the attention of activists, experts, and international organizations. As recent international efforts have failed systematically to provide measures that could help reduce global warning, we expect climate activism and climate and environmental migration, along other urgent issues, to intensify and get more attention from the LACEP literatures in the near future.

References

Abers, R., 2019. Bureaucratic Activism: Pursuing Environmentalism Inside the Brazilian State. *Latin American Politics and Society*. 61, 21–44.

Abers, R., Keck, M. E., 2013. *Practical Authority: Agency and Institutional Change in Brazilian Water Politics*. Oxford University Press, New York.

Abers, R., Keck, M. E., 2009. Mobilizing the State: The Erratic Partner in Brazil's Participatory Water Policy. *Politics and Society*. 37, 289–314.

Acselrad, H., 2010. Ambientalização das lutas sociais: o caso do movimento por justiça ambiental. *Estudos Avançados*. 24, 103–119.

Acselrad, H., 2008. Grassroots Reframing of Environmental Struggles in Brazil, in Carruthers, D. V. (Ed.), *Environmental Justice in Latin America: Problems, Promise, and Practice*. The MIT Press, Cambridge, MA, pp. 75–97.

Acselrad, H., 2006. Las políticas ambientales ante las coacciones de la globalización, in Alimonda, H. (Ed.), *Los tormentos de la materia: aportes para una ecología política latinoamericana*. CLACSO, Buenos Aires, pp.195–212.

Acselrad, H., 2004. *Conflitos ambientais no Brasil*. Relumé Dumará, Rio de Janeiro.

Aguilar-Støen, M., 2018. Social Forestry Movements and Science-Policy Networks: The Politics of the Forestry Incentives Program in Guatemala. *Geoforum*. 90, 20–26.

Aguilar-Støen, M., Hirsch, C., 2015. REDD+ and Forest Governance in Latin America: The Role of Science-Policy Networks, in Bull, B., Aguilar-Støen, M. (Eds.), *Environmental Politics in Latin America: Elite Dynamics, the Left Tide and Sustainable Development*. Routledge, Abingdon, pp. 171–189.

Ahmed, A., Johnson, M., Vásquez-Cortés, M., 2021. Land Titling, Race, and Political Violence: Theory and Evidence from Colombia, unpublished manuscript. https://cpb-us-w2.wpmucdn.com/campuspress.yale.edu/dist/8/3395/files/2020/06/AMM_Titling_Race_and_Violence.pdf

Alcañiz, I., 2016. *Environmental and Nuclear Networks in the Global South: How Skills Shape International Cooperation*. Cambridge University Press, New York.

Alcañiz, I., Berardo, R., 2016. A Network Analysis of Transboundary Water Cooperation in La Plata Basin. *Water Policy*. 18, 1120–1138.

Alcañiz, I., Gutiérrez, R. A., 2020a. Between the Global Commodity Boom and Subnational State Capacities: Payment for Environmental Services to Fight Deforestation in Argentina. *Global Environmental Politics*. 20, 38–59.

Alcañiz, I., Gutiérrez, R. A., 2020b. Gender, Land Distribution, and Who Gets State Funds to Stop Deforestation in Argentina. *Journal of Environmental Management*. 272, 1–8.

Alcañiz, I., Gutiérrez, R. A., 2009. From Local Protests to the International Court of Justice: Forging Environmental Foreign Policy in Argentina, in Harris, P. G. (Ed.), *Environmental Change and Foreign Policy*. Routledge Press, New York, pp. 109–120.

Alcañiz, I., Sanchez-Rivera, A. I., 2021. Climate Disasters, Inequality, and Perceptions of Government Assistance, in Sowers, J., VanDeveer, S. D., Weinthal, E. (Eds.), *Oxford Handbook of Comparative Environmental Politics*, Oxford University Press, New York.

Alimonda, H., 2008. Introducción, in Alvarado Merino, G., Delgado Ramos, G. C., Domínguez, D. et al. (Eds.), *Gestión ambiental y conflicto social en América Latina*. CLACSO, Buenos Aires, pp. 13–21.

Alimonda, H. (Ed.), 2006. *Los tormentos de la materia: aportes para una ecología política latinoamericana*. CLACSO, Buenos Aires.

Alonso, A., Costa, V., Maciel, D., 2008. Identity and Strategy in the Formation of the Brazilian Environmental Movement. *Novos Estudos-CEBRAP*. 4(SE). http://socialsciences.scielo.org/pdf/s_nec/v4nse/scs_a01.pdf

Alvarado Merino, G., 2008. Políticas neoliberales en el manejo de los recursos naturales en Perú: el caso del conflicto minero de Tambogrande, in Alvarado Merino, G., Delgado Ramos, G. C. et al. (Eds.), *Gestión ambiental y conflicto social en América Latina*. CLACSO, Buenos Aires, pp. 67–103.

Amengual, M., 2016. *Politicized Enforcement in Argentina: Labor and Environmental Regulation*. Cambridge University Press, New York.

Amengual, M., 2013. Pollution in the Garden of the Argentine Republic: Building State Capacity to Escape from Chaotic Regulation. *Politics and Society*. 41, 527–560.

Amengual, M., Dargent, E., 2020. The Social Determinants of Enforcement, in Brinks, D. M., Levitsky, S., Murillo, M. V. (Eds.), *The Politics of Institutional Weakness in Latin America*. Cambridge University Press, Cambridge, pp. 161–182.

Arce, M., 2014. *Resource Extraction and Protest in Peru*. University of Pittsburgh Press, Pittsburgh.

Arce, M., Miller, R. E., Patane, C. F., Polizzi, M. S., 2018. Resource Wealth, Democracy and Mobilisation. *Journal of Development Studies*. 54, 949–967.

Armesto, J. J., Smith-Ramirez, C., Rozzi, R., 2001. Conservation Strategies for Biodiversity and Indigenous People in Chilean Forest Ecosystems. *Journal of the Royal Society of New Zealand*. 31, 865–877.

Azócar, G., Sanhueza, R., Aguayo, M., Romero, H., Muñoz, M. D., 2005. Conflicts for Control of Mapuche-Pehuenche Land and Natural Resources in the Biobío Highlands, Chile. *Journal of Latin American Geography.* 4(2), 57–76.

Barandiaran, J., Rubiano-Galvis, S., 2019. An Empirical Study of EIA Litigation Involving Energy Facilities in Chile and Colombia. *Environmental Impact Assessment Review.* 79, 106311.

Barkin, D., Lemus, B., 2016. Local Solutions for Environmental Justice, in Castro, F. de, Hogenboom, B., Baud, M. (Eds.), *Environmental Governance in Latin America.* Palgrave Macmillan, London, pp. 257–286.

Bauer, C. J., 2015. Water Conflicts and Entrenched Governance Problems in Chile's Market Model. *Water Alternatives.* 8(2), 147–172.

Baver, S. L., Lynch, B. D. (Eds.), 2006a. *Beyond Sun and Sand: Caribbean Environmentalisms.* Rutgers University Press, New Brunswick.

Baver, S. L., Lynch, B. D., 2006b. The Political Ecology of Paradise, in Baver, S. L., Lynch, B. D. (Eds.), *Beyond Sun and Sand: Caribbean Environmentalisms.* Rutgers University Press, New Brunswick, pp. 3–16.

Bebbington, A., 2011. Elementos para una ecología política de los movimientos sociales y el desarrollo territorial en zonas mineras, in Bebbington, A. (Ed.), *Minería, movimientos sociales y respuestas campesinas: una ecología política de transformaciones territoriales.* IEP-CEPES, Lima, pp. 53–76.

Bebbington, A., Bury, J. (Eds.), 2013. *Subterranean Struggles: New Dynamics of Mining, Oil, and Gas in Latin America.* University of Texas Press, Austin.

Bebbington, A., Bury, J., Bebbington, D. H. et al., 2011. Movimientos sociales, lazos transnacionales y desarrollo territorial rural en zonas de influencia minera: Cajamarca-Perú y Cotacachi-Ecuador, in Bebbington, A. (Ed.), *Minería, movimientos sociales y respuestas campesinas: una ecología política de transformaciones territoriales.* IEP-CEPES, Lima, pp. 151–187.

Bledsoe, A., 2019. Afro-Brazilian Resistance to Extractivism in the Bay of Aratu. *Annals of the American Association of Geographers.* 109, 492–501.

Bolados García, P., Sánchez Cuevas, A., 2017. Una ecología política feminista en construcción: El caso de las" Mujeres de zonas de sacrificio en resistencia", Región de Valparaíso, Chile. *Psicoperspectivas.* 16, 33–42.

Bose, P., 2017. Land Tenure and Forest Rights of Rural and Indigenous Women in Latin America: Empirical Evidence. *Women's Studies International Forum.* 65, 1–8.

Bragagnolo, C., Lemos, C. C., Ladle, R. J., Pellin, A., 2017. Streamlining or Sidestepping? Political Pressure to Revise Environmental Licensing and EIA in Brazil. *Environmental Impact Assessment Review.* 65, 86–90.

Branfors, S., Borges, T., 2019. Brazil on the Precipice: From Environmental Leader to Despoiler (2010–2020), *Mongabay Series*: Amazon Conservation. https://news.mongabay.com/2019/12/brazil-on-the-precipice-from-environmental-leader-to-despoiler-2010-2020/

Bratman, E., 2015. Passive Revolution in the Green Economy: Activism and the Belo Monte Dam. *International Environmental Agreements*. 15, 61–77.

Börner, J., Wunder, S., 2008. Paying for Avoided Deforestation in the Brazilian Amazon: From Cost Assessment to Scheme Design. *International Forestry Review*. 10(3), 496–511.

Bull, B., Aguilar-Støen, M., 2015. Environmental Governance and Sustainable Development in Latin America, in Bull, B., Aguilar-Støen, M. (Eds.), *Environmental Politics in Latin America: Elite Dynamics, the Left Tide and Sustainable Development*. Routledge, Abingdon, pp. 1–14.

Burac, M., 2006. The Struggle for Sustainable Tourism in Martinique, in Baver, S. L., Lynch, B. D. (Eds.), *Beyond Sun and Sand: Caribbean Environmentalisms*. Rutgers University Press, New Brunswick and London, pp. 65–74.

Calvo, E., Murillo, M. V., 2019. *Non-Policy Politics: Richer Voters, Poorer Voters, and the Diversification of Electoral Strategies*. Cambridge University Press, Cambridge.

Calvo, E., Murillo, M. V., 2004. Who Delivers? Partisan Clients in the Argentine Electoral Market. *American Journal of Political Science*. 48, 742–757.

Carey, M., 2009. Latin American History: Current Trends, Interdisciplinary Insights, and Future Directions. *Environmental History*. 14, 221–252.

Carlin, R. E., Singer, M. M., Zechmeister, E. J., 2015. *The Latin American Voter: Pursuing Representation and Accountability in Challenging Contexts*. University of Michigan Press, Ann Arbor.

Carruthers, D., 2001. Environmental Politics in Chile: Legacies of Dictatorship and Democracy. *Third World Quarterly*. 22, 343–358.

Carruthers, D., Rodriguez, P., 2009. Mapuche Protest, Environmental Conflict and Social Movement Linkage in Chile. *Third World Quarterly*. 30, 743–760.

Carruthers, D. V. (Ed.), 2008a. *Environmental Justice in Latin America: Problems, Promise, and Practice*. The MIT Press, Cambridge, MA.

Carruthers, D. V., 2008b. Introduction: Popular Environmentalism and Social Justice in Latin America, in Carruthers, D. V. (Ed.), *Environmental Justice in Latin America: Problems, Promise, and Practice*. The MIT Press, Cambridge, MA, pp. 1–22.

Cartagena Cruz, R., 2017. Conflictos ambientales y movimientos sociales en El Salvador de la posguerra, in Almeida, P., Cordero Ulate, A. (Eds.), *Movimientos sociales en América Latina: perspectivas, tendencias y casos.* CLACSO, Buenos Aires, pp. 413–443.

Castro, F. de, Hogenboom, B., Baud, M. (Eds.), 2016. *Environmental Governance in Latin America.* Palgrave Macmillan, Basingstoke, Hampshire.

Ceddia, M. G., Gunter, U., Corriveau-Bourque, A., 2015. Land Tenure and Agricultural Expansion in Latin America: The Role of Indigenous Peoples' and Local Communities' Forest Rights. *Global Environmental Change.* 35, 316–322. https://doi.org/10.1016/j.gloenvcha.2015.09.010

Challenger, A., Córdova, A., Chavero, E. L., 2018. La opinión experta evalúa la política ambiental mexicana: Hacia la gestión de socioecosistemas. *Gestión Política Pública.* 43, 431–473.

Christel, L. G., 2020. Resistencias sociales y legislaciones mineras en las provincias argentinas: los casos de Mendoza, Córdoba, Catamarca y San Juan (2003–2009). *Política Gobierno.* XXVII, 3–24.

Christel, L. G., 2019. Derechos ambientales y resistencias sociales: el instrumento legal como repertorio contra la minería en Argentina. *Revista Austral Ciencias Sociales.* 36, 191–211.

Christel, L. G., Gutiérrez, R. A., 2021. The Political Impact of Environmental Mobilization: A Theoretical Discussion in the Light of the Argentine Case. *Canadian Journal of Latin American and Caribbean Studies/Revue canadienne des études latino-américaines et caraïbes.* 46 (1), 57–76.

Christel, L. G., Gutiérrez, R. A., 2017. Making Rights Come Alive: Environmental Rights and Modes of Participation in Argentina. *Journal of Environmental Development.* 26, 322–347.

Christel, L. G., Gutiérrez, R. A., 2023. Beyond the Lenses of Social Movements: Environmental Mobilization in Latin America, in Rossi, F. M. (Ed.), *The Oxford Handbook of Latin American Social Movements.* Oxford University Press, Oxford, pp. 1–15.

Christen, C., Herculano, S., Hochstetler, K. et al., 1998a. Latin American Environmentalism: Comparative Views. *Studies in Comparative International Development.* 32, 58–87.

Cisneros, P. (Ed.), 2016. *Política minera y sociedad civil en América Latina.* Editorial IAEN, Quito.

Cordero Ulate, A., 2017. Bosque, agua y lucha: movimientos ambientalistas en Costa Rica, in Almeida, P., Cordero Ulate, A. (Eds.), *Movimientos sociales en América Latina: perspectivas, tendencias y casos.* CLACSO, Buenos Aires, pp. 445–473.

da Rocha, D. F., Porto, M. F., Pacheco, T., Leroy, J. P., 2018. The Map of Conflicts Related to Environmental Injustice and Health in Brazil. *Sustainability Science*. 13, 709–719.

Delmas, M. A. and Young O. R. (Eds), 2009. *Governance for the Environment*. Cambridge University Press, New York.

De Pourcq, K., Thomas, E., Arts, B. et al., 2017. Understanding and Resolving Conflict between Local Communities and Conservation Authorities in Colombia. *World Development*. 93, 125–135.

Dixit, A., Londregan, J., 1996. The Determinants of Success of Special Interests in Redistributive Politics. *Journal of Politics*. 58, 1132–1155.

Dobson, A., 1998. *Justice and the Environment: Conceptions of Environmental Sustainability and Theories of Distributive Justice*. Oxford University Press, New York.

Domínguez, D., 2008. La transhumancia de los campesinos kollas: ¿hacia un modelo de desarrollo sustentable?, in Alvarado Merino, G., Delgado Ramos, G. C., Domínguez, D. et al. (Eds.), *Gestión ambiental y conflicto social en América Latina*. CLACSO, Buenos Aires, pp. 137–191.

Durazo, E. P., 2004. Política y gestión ambiental contemporánea en México. *Economics Information*. 328, 5–24.

Eakin, H., Lemos, M. C., 2006. Adaptation and the State: Latin America and the Challenge of Capacity-Building under Globalization. *Global Environmental Change*. 16, 7–18.

Ebeling, J., Yasué, M., 2009. The Effectiveness of Market-Based Conservation in the Tropics: Forest Certification in Ecuador and Bolivia. *Journal of Environmental Management*. 90, 1145–1153.

Edwards, G., Roberts, J. T., 2015. *A Fragmented Continent: Latin America and the Global Politics of Climate Change*. The MIT Press, Cambridge, MA.

Falleti, T. G., Riofrancos, T. N., 2018. Endogenous Participation: Strengthening Prior Consultation in Extractive Economies. *World Politics*. 70(1), 86–121.

Fearnside, P. M., 2003. Conservation Policy in Brazilian Amazonia: Understanding the Dilemmas. *World Development*. 31, 757–779.

Fernández-Milmanda, B., Garay, C., 2020. A Multilevel Approach to Enforcement, in Daniel M. Brinks, Steven Levitsky and María Victoria Murillo (Eds.), *The Politics of Institutional Weakness in Latin America*. Cambridge University Press, Cambridge, pp. 183–207.

Fernández-Milmanda, B., Garay, C., 2019. Subnational Variation in Forest Protection in the Argentine Chaco. *World Development*. 118, 79–90.

Fearnside, P. M., 2017. How a Dam Building Boom is Transforming the Brazilian Amazon. *Yale Environment*, 360. https://e360.yale.edu/features/how-a-dam-building-boom-is-transforming-the-brazilian-amazon

Foyer, J., Dumoulin Kevran, D., 2017. ¿Ambientalismo de las ONG versus ambientalismo de los pobres? in Almeida, P., Cordero Ulate, A. (Eds.), *Movimientos sociales en América Latina: perspectivas, tendencias y casos.* CLACSO, Buenos Aires, pp. 391–412.

Freire, G. N., Schwartz Orellana, S. D., Zumaeta Aurazo, M. et al., 2015. *Indigenous Latin America in the Twenty-First Century: The First Decade* (No. 98544, pp. 1–120). The World Bank, Washington, DC.

Futemma, C., Munari, L. C., Adams, C., 2015. The Afro-Brazilian Collective Land: Analyzing Institutional Changes in the Past Two Hundred Years. *Latin American Research Review.* 50(4), 26–48.

Gagnon-Légaré, A., Prestre, P. L., 2014. Explaining Variations in the Subnational Implementation of Global Agreements: The Case of Ecuador and the Convention on Biological Diversity. *The Journal of Environment & Development.* 23, 220–246.

García-López, G. A., Arizpe, N., 2010. Participatory Processes in the Soy Conflicts in Paraguay and Argentina. *Ecological Economics.* 70, 196–206.

Gerlak, A. K., Saguier, M., Mills-Novoa, M. et al., 2020. Dams, Chinese Investments, and EIAs: A Race to the Bottom in South America? *Ambio.* 49, 156–164.

Gizelis, T., Wooden, A., 2010. Water Resources, Institutions, & Intrastate Conflict. *Political Geography.* 29, 444–453.

Göbel, B., Ulloa, A. (Eds.), 2014. *Colombia y el extractivismo en América Latina.* Universidad Nacional de Colombia, Bogotá.

Gonzalez, A., 2021. Voiceless Development, Toxic Injustice, Criminal Resistance: A Study of Peruvian Natural Resource Extraction Through the Political Ecology of Voice, in Ioris, A. A. R. (Ed.), *Environment and Development.* Palgrave Macmillan, Cham, pp. 305–335.

Gudynas, E., 2009. La ecología política del giro biocéntrico en la nueva Constitución de Ecuador. *Revista de Estudios Sociales.* (32), 34–47.

Guevara Sanginés, A. E., 2005. Política ambiental en México: génesis, desarrollo y perspectivas. *Third World Quarterly.* 38, 1146–1163.

Gustafsson, M. T., 2017. The Struggles Surrounding Ecological and Economic Zoning in Peru. *Third World Quarterly.* 38, 1146–1163.

Gutiérrez, R. A., 2020. A Troubled Collaboration: Cartoneros and the PRO Administrations in Buenos Aires. *Latin American Politics and Society.* 62, 97–120.

Gutiérrez, R. A. (Ed.), 2018. *Construir el ambiente: sociedad, estado y políticas ambientales en Argentina.* Teseo, Buenos Aires.

Gutiérrez, R. A., 2017. La confrontación de coaliciones sociedad-estado: la política de protección de bosques nativos en Argentina (2004–2015). *Rev. SAAP* 11, 283–312.

Gutiérrez, R. A., 2010. When Experts Do Politics: Introducing Water Policy Reform in Brazil. *Governance.* 23, 59–88.

Gutiérrez, R. A., Isuani, F. J., 2014. La emergencia del ambientalismo estatal y social en Argentina. *Revista de Administração Pública.* 48, 295–322.

Gutiérrez, R., Jones, A., 2004. Corporate Social Responsibility in Latin America: An Overview of Its Characteristics and Effects on Local Communities, in Cotreras, M. (Ed.), *Corporate Social Responsibility in Asia and Latin America.* Inter-American Development Bank, Whasington, DC, pp. 151–187.

Habermas, J., 1996. *Between Facts and Norms: Contributions to a Discourse Theory of Law and Democracy.* The MIT Press, Cambridge, MA.

Harrison, K., 1996. *Passing the Buck: Federalism and Canadian Environmental Policy.* University of British Columbia Press, Vancouver.

Herrera, V., 2017. *Water and Politics: Clientelism and Reform in Urban Mexico.* University of Michigan Press, Ann Arbor.

Herrera, V., Mayka, L., 2020. How Do Legal Strategies Advance Social Accountability? Evaluating Mechanisms in Colombia. *The Journal of Development Studies.* 56, 1437–1454.

Hochstetler, K., 2021. Climate Institutions in Brazil: Three Decades of Building and Dismantling Climate Capacity. *Environmental Politics.* 30, 49–70.

Hochstetler, K., 2020. *Political Economies of Energy Transition: Wind and Solar Power in Brazil and South Africa.* Cambridge University Press, Cambridge, MA.

Hochstetler, K., Keck, M. E., 2007. *Greening Brazil: Environmental Activism in State and Society.* Duke University Press, Durham.

Hochstetler, K., Kostka, G., 2015. Wind and Solar Power in Brazil and China: Interests, State–Business Relations, and Policy Outcomes. *Global Environmental Politics.* 15, 74–94.

Hochstetler, K., Viola, E., 2012. Brazil and the Politics of Climate Change: Beyond the Global Commons. *Environmental Politics.* 21, 753–771.

Hogenboom, B., 2015. New Elites around South America's Strategic Resources, in Bull, B., Aguilar-Støen, M. (Eds.), *Environmental Politics in Latin America: Elite Dynamics, the Left Tide and Sustainable Development.* Routledge, Abingdon, pp. 113–130.

Hornborg, A., 2009. Zero-Sum World: Challenges in Conceptualizing Environmental Load Displacement and Ecologically Unequal Exchange in the World System. *International Journal of Comparative Sociology.* 50, 237–262.

Inoue, C. Y. A., Moreira, P. F., 2017. Many Worlds, Many Nature (s), One Planet: Indigenous Knowledge in the Anthropocene. *Revista Brasileira de Política Internacional.* 59(2), 1–19.

Jácome, F., 2006. Environmental Movements in the Caribbean, in Baver, S. L., Lynch, B. D. (Eds.), *Beyond Sun and Sand: Caribbean Environmentalisms.* Rutgers University Press, New Brunswick, pp. 17–31.

Kasara, K., Suryanarayan, P., 2015. When Do the Rich Vote Less than the Poor and Why? Explaining Turnout Inequality across the World. *American Journal of Political Science.* 59, 613–627.

Kauffman, C. M., Martin, P. L., 2017. Can Rights of Nature Make Development More Sustainable? Why Some Ecuadorian Lawsuits Succeed and Others Fail. *World Development.* 92, 130–142.

Kauffman, C. M., Terry, W., 2016. Pursuing Costly Reform: The Case of Ecuadorian Natural Resource Management. *Latin American Research Review.* 51, 163–185.

Klepek, J., 2012. Against the Grain: Knowledge Alliances and Resistance to Agricultural Biotechnology in Guatemala. *Canadian Journal of Development Studies/Revue Canadienne D'études du Développement.* 33, 310–325.

Kohl, B., Farthing, L., 2012. Material Constraints to Popular Imaginaries: The Extractive Economy and Resource Nationalism in Bolivia. *Political Geography.* 31, 225–235.

Lasso, M., 2019. *Erased: The Untold Story of the Panama Canal.* Harvard University Press, Cambridge, MA.

Leguizamón, A., 2020. *Seeds of Power: Environmental Injustice and Genetically Modified Soybeans in Argentina.* Duke University Press, Durham.

Leite,J. R. M., Venâncio,M. D., 2017. A Proteção Ambiental no Superior Tribunal de Justiça: protegendo o meio ambiente por intermédio da operacionalização do Estado de Direito Ecológico. *Sequência.* 77, 29–50.

Lemos, M. C. de M., 1998. The Politics of Pollution Control in Brazil: State Actors and Social Movements Cleaning up Cubatão. *World Development.* 26, 75–87.

Lemos, M. C.de M., Looye, J., 2003. Looking for Sustainability: Environmental Coalitions across the State-Society Divide. *Bulletin of Latin American Research.* 22, 350–370.

Lewis, T. L., 2016. *Ecuador's Environmental Revolutions: Ecoimperialists, Ecodependents, and Ecoresisters.* MIT Press, Cambridge, MA.

Luna, J. P., 2010. Segmented Party-Voter Linkages in Latin America: The Case of the UDI. *Journal of Latin American Studies*. 42, 325–356.

Lynch, B. D., 2006. Conclusion: Toward a Creole Environmentalism, in Baver, S. L., Lynch, B. D. (Eds.), *Beyond Sun and Sand: Caribbean Environmentalisms*. Rutgers University Press, New Brunswick, pp. 158–170.

MacLeod, M., Park, J., 2011. Financial Activism and Global Climate Change: The Rise of Investor-Driven Governance Networks. *Global Environmental Politics*. 11(2), 54–74.

Madariaga, A., Allain, M., 2020. Contingent Coalitions in Environmental Policymaking: How Civil Society Organizations Influenced the Chilean Renewable Energy Boom. *Policy Studies Journal*. 48(3), 672–699.

Mahlknecht, J., González-Bravo, R., Loge, F. J., 2020. Water-Energy-Food Security: A Nexus Perspective of the Current Situation in Latin America and the Caribbean. *Energy*. 194, 116824.

Martínez-Alier, J., 2008. Conflictos ecológicos y justicia ambiental. *Papeles de relaciones ecosociales y cambio global*. (103), 11–27.

Martínez-Alier, J., 2007. *O ecologismo dos pobres*. Editora Contexto, São Paulo.

Martínez-Alier, J., 2004. Ecological Distribution Conflicts and Indicators of Sustainability. *International Journal of Political Economy*. 34, 13–30.

Martínez-Alier, J., Baud, M., Sejenovich, H., 2016. Origins and Perspectives of Latin American Environmentalism, in Castro, F. de, Hogenboom, B., Baud, M. (Eds.), *Environmental Governance in Latin America*. Palgrave Macmillan, Basingstoke, pp. 26–57.

Martínez-Alier, J., Kallis, G., Veuthey, S., Walter, M., Temper, L., 2010. Social Metabolism, Ecological Distribution Conflicts, and Valuation Languages. *Ecological Economics*. 70, 153–158.

Martínez-Alier, J., Séjenovich, H., Baud, M., 2015. El ambientalismo y ecologismo latinoamericano, in Castro, F. de, Hogenboom, B., Baud, M. (Eds.), *Gobernanza ambiental en América Latina*. CLACSO, Buenos Aires, pp. 39–72.

McAllister, L. K., 2008. *Making Law Matter: Environmental Protection and Legal Institutions in Brazil*. Stanford University Press, Stanford.

McCaffrey, K. T., Baver, S. L., 2006. "Ni Una Bomba Más": Reframing the Vieques Struggle, in Baver, S. L., Lynch, B. D. (Eds.), *Beyond Sun and Sand: Caribbean Environmentalisms*. Rutgers University Press, New Brunswick, pp. 109–128.

Merlinsky, G., 2013a. *Política, derechos y justicia ambiental: el conflicto del Riachuelo*. Fondo de Cultura Económica, Buenos Aires.

Merlinsky, G. (Ed.), 2013b. *Cartografías del conflicto ambiental en Argentina.* Ediciones Ciccus, Buenos Aires.

Merlinsky, G., 2013c. La cuestión ambiental en la agenda pública, in Merlinsky, G. (Ed.), *Cartografías del conflicto ambiental en Argentina.* Ediciones Ciccus, Buenos Aires, pp. 19–60.

Micheli, J., 2002. Política ambiental en México y su dimensión regional. *Región y Sociedad.* 14, 129–170.

Mollett, S., 2014. A Modern Paradise: Garifuna Land, Labor, and Displacement-in-Place. *Latin American Perspectives.* 41, 27–45.

Mumme, S. P., 2007. Trade Integration, Neoliberal Reform, and Environmental Protection in Mexico: Lessons for the Americas. *Latin American Perspectives.* 34, 91–107.

Murillo, M. V., Mangonnet, J., 2013. La economía política de la Argentina exportadora en el nuevo milenio: proponiendo una nueva agenda de investigación. *Desarrollo Económico.* 53, 223–239.

Murphy, D. F., Bendell, J., 1997. *In the company of partners: business, environmental groups and sustainable development post-Rio.* Policy Press, Bristol.

Narváez, I., 2007. La política ambiental del Estado: ¿Hacia el colapso del modelo de conservación? in Fontaine, G., Narváez, I. (Eds.), *Yasuní en el siglo XXI: el Estado ecuatoriano y la conservación de la Amazonía,* FLACSO Ecuador, Quito, pp. 33–73.

Nicolle, S., Leroy, M., 2017. Advocacy Coalitions and Protected Areas Creation Process: Case Study in the Amazon. *Journal of Environmental Management.* 198, 99–109.

O'Connor, J., 1998. *Natural Causes: Essays in Ecological Marxism.* The Guilford Press, New York.

Offe, C., 1992. *Partidos políticos y nuevos movimientos sociales.* Editorial Sistema, Madrid.

Oliveros, V., 2018. "The Squeaky Wheel Gets the Grease"? The Conflict Imperative and the Slow Fight against Environmental Injustice in Northern Peruvian Amazon. *Ecology and Society,* 23(3), 7–19.

Orta-Martínez, M., Pellegrini, L., Arsel, M., 2018. The Conflict Imperative and the Slow Fight against Environmental Injustice in Northern Peruvian Amazon. *Ecology and Society.* 23, art7.

Pacheco-Vega, R., 2020. Governing Urban Water Conflict through Watershed Councils—A Public Policy Analysis Approach and Critique. *Water.* 12, 1849.

Pacheco-Vega, R., 2019. (Re) theorizing the Politics of Bottled Water: Water Insecurity in the Context of Weak Regulatory Regimes. *Water.* 11, 658.

Paredes, M., 2018. Transnational Advocacy and Local State Capacity: The peruvian ombuds Office and the Protection of Indigenous Rights, in Evans, P., Rodríguez Garavito, C. (Eds.), *Transnational Networks Advocacy: Twenty Years of Evolving Theory and Practice*, Dejusticia, Bogotá, pp. 92–108.

Paredes, M., 2006. Discurso indígena y conflicto minero en el Perú., in Iguíniz, J., Escobal, J., Degregori, C. I. (Eds.), *Perú: el problema agrario en debate*. SEPIA XI – Seminario Permanente de Investigación Agrícola. SEPIA, Lima, pp. 501–539.

Paredes, M., Kaulard, A., 2020. Fighting the Climate Crisis in Persistently Unequal Land Regimes: Natural Protected Areas in the Peruvian Amazon. *Journal of Cleaner Production*. 265, 1–9.

Pellegrini, L., Arsel, M., 2018. Oil and Conflict in the Ecuadorian Amazon: An Exploration of Motives and Objectives. *European Review of Latin American and Caribbean Studies/Revista Europea de Estudios Latinoamericanos y del Caribe*. (106), 217–226.

Pereira, J. C., Viola, E., 2019. Catastrophic Climate Risk and Brazilian Amazonian Politics and Policies: A New Research Agenda. *Global Environmental Politics*. 19, 93–103.

Perez Guartambel, C., 2006. *Justicia indígena*. Universidad de Cuenca, Cuenca.

Perry, 2009. "If We Didn't Have Water": Black Women's Struggle for Urban Land Rights in Brazil. *Environ. Justice*. 2, 9–14.

Pielke Jr., R. A., 2007. *The Honest Broker: Making Sense of Science in Policy and Politics*. Cambridge University Press, New York.

Polanyi, K., 2001. *The Great Transformation: The Political and Economic Origins of Our Time*. Beacon Press, New York.

Ponce, A., McClintock, C., 2014. The Explosive Combination of Inefficient Local Bureaucracies and Mining Production: Evidence from Localized Societal Protests in Peru. *Latin American Politics and Society*. 53, 118–140.

Pragier, D., 2019. Comunidades indígenas frente a la explotación de litio en sus territorios: contextos similares, respuestas distintas. *Polis*. 52, 76–91.

Riofrancos, T., 2020. *Resource Radicals: From Petro-Nationalism to Post-Extractivism in Ecuador*. Duke University Press, Durham.

Robin S., Pinedo-Vasquez, M., 2011. Forest Policy Reform and the Organization of Logging in Peruvian Amazonia. *Development and Change*. 42, 609–631.

Rodríguez, I., Robledo, J., Sarti, C., Borel, R., Melace, A. C., 2015. Abordando la Justicia Ambiental desde la transformación de conflictos: experiencias con Pueblos Indígenas en América Latina. *Revista de Paz y Conflictos*. 8, 97–128.

Rodriguez-Díaz, C. E., Lewellen-Williams, C., 2020. Race and Racism as Structural Determinants for Emergency and Recovery Response in the Aftermath of Hurricanes Irma and Maria in Puerto Rico. *Health Equity.* 4, 232–238.

Sachs, W., 1999. Sustainable Development and the Crisis of Nature: On the Political Anatomy of an Oxymoron, in Fischer, F., Hajer, M. A. (Eds.), *Living With Nature: Environmental Politics as Cultural Discourse.* Oxford University Press, New York, pp. 23–41.

Scheberle, D., 2005. The Evolving Matrix of Environmental Federalism and Intergovernmental Relationships. *Publius: The Journal of Federalism.* 35, 69–86.

Sedrez, L., 2009. Latin American Environmental History: A Shifting Old/New Field, in Burke, E., Pomeranz, K. (Eds.), *The Environment and World History.* University of California Press, California, pp. 255–275.

Sempertegui, A., 2021. Indigenous Women's Activism, Ecofeminism, and Extractivism: Partial Connections in the Ecuadorian Amazon. *Politics & Gender.* 17, 197–224.

Silva, E., 2016. Afterword: From Sustainable Development to Environmental Governance, in Castro, F. de, Hogenboom, B., Baud, M. (Eds.), *Environmental Governance in Latin America.* Palgrave Macmillan, Basingstoke, pp. 326–335.

Silva, E., Akchurin, M., Bebbington, A. J., 2018. Policy Effects of Resistance against Mega-Projects in Latin America: An Introduction. *European Review of Latin American and Caribbean Studies/Revista Europea de Estudios Latinoamericanos y del Caribe.* (106), 27–47.

Silva, E., Kaimowitz, D., Bojanic, A. et al., 2002. Making the Law of the Jungle: The Reform of Forest Legislation in Bolivia, Cameroon, Costa Rica, and Indonesia. *Global Environmental Politics.* 2, 63–97.

Steinberg, P. F., 2001. *Environmental Leadership in Developing Countries: Transnational Relations and Biodiversity Policy in Costa Rica and Bolivia.* MIT Press, Cambridge, MA.

Stokes, S. C., 2005. Perverse Accountability: A Formal Model of Machine Politics with Evidence from Argentina. *American Political Science Review.* 99(3), 315–325.

Sundberg, J., 2008. Placing Race in Environmental Justice Research in Latin America. *Society and Natural Resources.* 21, 569–582.

Svampa, M., 2019. *Neo-Extractivism in Latin America: Socio-Enviornmental Conflicts, the Territorial Turn, and the New Political Narratives.* Cambridge University Press, Cambridge.

Svampa, M., 2012. Consenso de los commodities, giro ecoterritorial y pensamiento crítico en América Latina. Observatorio Social de **América** Latina. XIII, 15–39.

Szwarcberg, M., 2015. *Mobilizing Poor Voters: Machine Politics, Clientelism, and Social Networks in Argentina* (Vol. 38). Cambridge University Press, Cambridge.

Tecklin, D., Bauer, C., Prieto, M., 2011. Making Environmental Law for the Market: The Emergence, Character, and Implications of Chile's Environmental Regime. *Environmental Politics*. 20, 879–898.

Tigre, M. A., 2019. Building a Regional Adaptation Strategy for Amazon Countries. *International Environmental Agreements: Politics, Law and Economics*. 19, 411–427.

Tormos-Aponte, F., 2018. The Politics of Survival in Puerto Rico: The Balance of Forces in the Wake of Hurricane María. *Alternautas*. 5(2), 79–94.

Torres Ramirez, B., 2019. La participación de méxico en la convención marco de las naciones unidas sobre el cambio climático. *Foro Internacional*. LIX, 1179–1219.

Ulloa, A., 2017. Perspectives of Environmental Justice from Indigenous Peoples of Latin America: A Relational Indigenous Environmental Justice. *Environmental Justice*. 10, 175–180.

Urkidi, L., 2010. A Global Environmental Movement against Gold Mining: Pascua–Lama in Chile. *Ecol. Econ.* 70, 219–227.

Urkidi, L., Walter, M., 2011. Dimensions of Environmental Justice in Anti-Gold Mining Movements in Latin America. *Geoforum*. 42, 683–695.

Valdés Pizzini, M., 2006. Historical Contentions and Future Trends in the Coastal Zones: The Environmental Movement in Puerto Rico, in Baver, S. L., Lynch, B. D. (Eds.), *Beyond Sun and Sand: Caribbean Environmentalisms*. Rutgers University Press, New Brunswick, pp. 44–64.

Valladares, C., Boelens, R., 2017. Extractivism and the Rights of Nature: Governmentality, "Convenient Communities" and Epistemic Pacts in Ecuador. *Environmental Politics*. 26, 1015–1034.

Velásquez Runk, J., 2012. Indigenous Land and Environmental Conflicts in Panama: Neoliberal Multiculturalism, Changing Legislation, and Human Rights. *J. Lat. Am. Geogr.* 11, 21–47.

Velázquez López Velarde, R., Somuano Ventura, M. F., Ortega Ortiz, R. Y., 2018. David contra Goliat: ¿Cómo los movimientos ambientalistas se enfrentan a las grandes corporaciones? *América Latina Hoy*. 79, 41–58.

Vélez, M. A., Robalino, J., Cardenas, J. C., Paz, A., Pacay, E., 2020. Is Collective Titling Enough to Protect Forests? Evidence from Afro-descendant Communities in the Colombian Pacific region. *World Development*. 128, 104837.

Vida, M., 2020. Central America's Endangered Lungs: On Nicaragua's Caribbean Coast, Black and Indigenous Forest Rangers Take Environmental

Protection into their Own Hands. *NACLA Report on the Americas*. 52, 214–226.

Viola, E. J., 1992. O movimento ambientalista no Brasil (1971–1991): da denúncia e conscientização pública para a institucionalização e o desenvolvimento sustentável, in Golbenberg, M. (Ed.), *Ecologia, ciência e política*. Revan, Rio de Janeiro.

Viola, E., Franchini, M., 2017. *Brazil and Climate Change: Beyond the Amazon*. Routledge, New York.

Viola, E., Franchini, M., 2014. Brazilian Climate Politics 2005–2012: Ambivalence and Paradox. *Wiley Interdisciplinary Reviews: Climate Change*. 5(5), 677–688.

Walter, M., Martínez-Alier, J., 2010. How to Be Heard When Nobody Wants to Listen: Community Action against Mining in Argentina. *Canadian Journal of Development Studies*. 30, 281–301.

Walter, M., Urkidi, L., 2016. Community Consultations and Large-Scale Mining in Latin America, in Castro, F. de, Hogenboom, B., Baud, M. (Eds.), *Environmental Governance in Latin America*. Palgrave Macmillan, Basingstoke, Hampshire, pp. 287–325.

Whittemore, M. E., 2011. The Problem of Enforcing Nature's Rights under Ecuador's Constitution: Why the 2008 Environmental Amendments Have No Bite. *Pacific Rim Law & Policy Journal*, 20(3), 659–691.

Wismer, S., Lopez de Alba Gomez, A., 2011. Evaluating the Mexican Federal District's Integrated Solid Waste Management Programme. *Waste Management & Research: The Journal for a Sustainable Circular Economy*. 29, 480–490.

Wong, M. T., 2018. *Natural Resources, Extraction and Indigenous Rights in Latin America: Exploring the Boundaries of Environmental and State-Corporate Crime in Bolivia, Peru and Mexico*. Routledge, London.

Worster, D., 1977. *Nature's Economy: The Roots of Ecology*. Sierra Club Books, San Francisco.

Zaremberg, G. and Wong, M. T., 2018. Participation on the edge: Prior consultation and extractivism in Latin America. *Journal of Politics in Latin America*, 10(3), pp. 29–58.

Zegarra, Eduardo, Orihuela, José Carlos, Paredes, Maritza. 2007. *Minería y economía de los hogares en la sierra peruana: impactos y espacios de conflicto*. Lima: GRADE. 85 p. Documento de trabajo, 51. www.grade.org .pe/wp-content/uploads/ddt51.pdf

Zhouri, A., 2010. "Adverse Forces" in the Brazilian Amazon: Developmentalism Versus Environmentalism and Indigenous Rights. *The Journal of Environment & Development*. 19(3), 252–273.

Zimmerer, K. S., 2011. Conservation Boom with Agricultural Growth? Sustainability and Shifting Environmental Governance in Latin America, 1985-2008 (Mexico, Costa Rica, Brazil, Peru, Bolivia). *Latin American Research Review*. 46, 82–114.

Zuluaga-Sánchez, G. P., Arango-Vargas, C., 2013. Mujeres campesinas: resistencia, organización y agroecología en medio del conflicto armado. *Cuadernos de Desarrollo Rural*. 10, 159–180.

Acknowledgments

We are grateful to the strong community of colleagues, friends, and family that sustained us during the process of writing this book. The normal challenges and difficulties of writing a book were exacerbated by the very abnormal times brought on by the global pandemic. Our network of support ensured that the writing process and these past two years were more meaningful, stimulating, and enjoyable than they otherwise would have been. We are also thankful to the editors of the Politics and Society in Latin America Elements series, Tulia Falleti, María Victoria Murillo, Juan Pablo Luna, and Andrew Schrank and two anonymous reviewers. For their institutional support, we thank the Latin American and Caribbean Studies Center (LACS), the department of Government and Politics, the College of Behavioral and Social Sciences of the University of Maryland, the National University of San Martín's Environment and Politics Area, and the University of Columbia's Institute of Latin American Studies. For research assistance, we thank Elin Berlin. The intellectual endeavor of writing this book benefited enormously from the expertise, insights, and friendship of many colleagues: Ernesto Calvo, Lucas Christel, Agustina Giraudy, Fernando Isuani, Marccus Hendricks, Kathryn Hochstetler, Alejandra Marchevsky, and Ana Sanchez-Rivera. We also thank Margaret Keck, extraordinary mentor and graduate advisor to Ricardo and one of few political scientists studying Latin America to sound an early alarm on the deterioration of the natural environment and reveal the importance of political and social alliances to stop it. Finally, we are especially indebted to our families, Ana Filippa, Ignacio Gutiérrez, Camila Gutiérrez and Ernesto Calvo, Camilo Calvo Alcañiz, Lauren Michelle, and Emiliano Calvo Alcañiz, who make our lives better.

*A las y los defensor@s de la justicia ambiental en Latinoamérica,
el Caribe, y más allá.*

Cambridge Elements ≡

Politics and Society in Latin America

Maria Victoria Murillo
Columbia University Maria

Victoria Murillo is Professor of Political Science and International Affairs at Columbia University. She is the author of Political Competition, Partisanship, and Policymaking in the Reform of Latin American Public Utilities (Cambridge, 2009). She is also editor of Carreras Magisteriales, Desempeño Educativo y Sindicatos de Maestros en América Latina (2003), and co-editor of Argentine Democracy: the Politics of Institutional Weakness (2005). She has published in edited volumes as well as in the American Journal of Political Science, World Politics, and Comparative Political Studies, among others.

Tulia G. Falleti
University of Pennsylvania

Tulia G. Falleti is the Class of 1965 Endowed Term Professor of Political Science, Director of the Latin American and Latino Studies Program, and Senior Fellow of the Leonard Davis Institute for Health Economics at the University of Pennsylvania. She received her BA in Sociology from the Universidad de Buenos Aires and her Ph.D. in Political Science from Northwestern University. Falleti is the author of Decentralization and Subnational Politics in Latin America (Cambridge University Press, 2010), which earned the Donna Lee Van Cott Award for best book on political institutions from the Latin American Studies Association, and with Santiago Cunial of Participation in Social Policy: Public Health in Comparative Perspective (Cambridge University Press, 2018). She is co-editor, with Orfeo Fioretos and Adam Sheingate, of The Oxford Handbook of Historical Institutionalism (Oxford University Press, 2016), among other edited books. Her articles on decentralization, federalism, authoritarianism, and qualitative methods have appeared in edited volumes and journals such as the American Political Science Review, Comparative Political Studies, Publius, Studies in Comparative International Development, and Qualitative Sociology, among others.

Juan Pablo Luna
The Pontifical Catholic University of Chile

Juan Pablo Luna is Professor of Political Science at The Pontifical Catholic University of Chile. He received his BA in Applied Social Sciences from the UCUDAL (Uruguay) and his PhD in Political Science from the University of North Carolina at Chapel Hill. He is the author of Segmented Representation. Political Party Strategies in Unequal Democracies (Oxford University Press, 2014), and has co-authored Latin American Party Systems (Cambridge University Press, 2010). In 2014, along with Cristobal Rovira, he co-edited The Resilience of the Latin American Right (Johns Hopkins University). His work on political representation, state capacity, and organized crime has appeared in the following journals: Comparative Political Studies, Revista de Ciencia Política, the Journal of Latin American Studies, Latin American Politics and Society, Studies in Comparative International Development, Política y Gobierno, Democratization, Perfiles Latinoamericanos, and the Journal of Democracy.

About the Series

Latin American politics and society are at a crossroads, simultaneously confronting serious challenges and remarkable opportunities that are likely to be shaped by formal institutions and informal practices alike. The Elements series on Politics and Society in Latin America offers multidisciplinary and methodologically pluralist contributions on the most important topics and problems confronted by the region.

Cambridge Elements ☰

Politics and Society in Latin America

A full series listing is available at: www.cambridge.org/PSLT